Good Thinking

Why Flawed Logic Puts Us All
at Risk and How Critical Thinking
Can Save the World

DAVID ROBERT GRIMES

THE EXPERIMENT
NEW YORK

GOOD THINKING: *Why Flawed Logic Puts Us All at Risk and How Critical Thinking Can Save the World*
Copyright © 2019 by David Robert Grimes

Originally published in the UK as *The Irrational Ape* by Simon & Schuster UK in 2019. First published in North America in revised form by The Experiment, LLC, in 2021.

The Experiment, LLC
220 East 23rd Street, Suite 600
New York, NY 10010-4658
theexperimentpublishing.com

THE EXPERIMENT and its colophon are registered trademarks of The Experiment, LLC. Many of the designations used by manufacturers and sellers to distinguish their products are claimed as trademarks. Where those designations appear in this book and The Experiment was aware of a trademark claim, the designations have been capitalized.

The Experiment's books are available at special discounts when purchased in bulk for premiums and sales promotions as well as for fund-raising or educational use. For details, contact us at info@theexperimentpublishing.com.

Library of Congress Cataloging-in-Publication Data available upon request

ISBN 978-1-61519-793-4
Ebook ISBN 978-1-61519-794-1

Cover design by Beth Bugler
Text design by M Rules
Author photograph by Abe Neihum

Manufactured in the United States of America

First printing March 2021
10 9 8 7 6 5 4 3 2 1

To Mathilde, Danny, and Laura—
for the inspiration, the ideas,
and the encouragement

CONTENTS

PART I: Without Reason
Formal Fallacies and How to Defeat Them

PART 2: The Pure and Simple Truth?
Spotting and Debunking Dubious Rhetoric

PART 3: Trapdoors of the Mind
The Struggle Between Reason and Belief

PART 4: Lies, Damned Lies, and Statistics
How Numbers Can Mislead Us

PART 5: News of the World
How Media Indulges Bad Thinking

PART 6: The Candle in the Dark
What Science Is and What It Is Not

When future historians come to chronicle the early 21st century, one curious paradox may epitomize the era in which we now live. The advent of the internet in the final decade of the 20th century promised the entire repository of the world's information at our very fingertips, and a new dawn of mutual understanding that would transcend the confines of geography or politics. Fewer sages foresaw, however, that the same technology would enable the propagation of falsehood at staggering velocity to huge, receptive audiences. Fewer still predicted this would leave us more polarized and divided than ever before. The upheaval and division that define our current epoch will no doubt intrigue those future historians. Navigating our way now, without the benefit of hindsight, is an abiding challenge. With such a confluence of contradictory aspects, it is small wonder we're frequently left feeling overwhelmed, and an ostensible old Chinese curse is frequently recited: "May you live in interesting times."

Fittingly perhaps in this age of disinformation, this invocation is apocryphal—there is no such idiom in Chinese. It's a benign misconception, but the unsettling reality is that many of the delusions we harbor can cause us serious harm. As I write this in late 2020, we are in the midst of the COVID-19 crisis. In this first pandemic of the information age, we've witnessed the emergence of a shadow plague: an "infodemic," where facts, half-truths, and outright fabrications haphazardly entangle, virulently spreading as much as the pathogen that created the havoc in the first place. In a vacuum of uncertainty, rumors and myths move aggressively to fill the void. In the sound and fury, we have seen a dark renaissance of conspiracy theory, propaganda, and disinformation exert a worrying influence—about 30 percent of the US population, for example, believe COVID-19 to be a hoax or deliberately engineered, despite the ample evidence to the contrary.

Inevitably perhaps, these bogus claims have become serious impediments to societal cohesion and collective well-being. Something as simple as face masks have become politicized. Around the world, protests against their apparent imposition have become a unifying flashpoint for something more fundamental: a focal point of contention. Beyond this superficial edifice lies a litany of conspiracy theories—that the virus is a hoax, or caused by 5G, or is an odious plot by Bill Gates and some sinister cabal. The astute observer too would notice the heavy involvement of the anti-vaccine community, alternative health advocates, and even supporters of QAnon, the outlandish far-right conspiracy theory that insists a secret organization of Satan-worshipping pedophiles runs the world. The stark reality is that the enduring legacy of the COVID era may not be solely the pandemic but a dawning awareness of quite how susceptible we are to poisonous fictions.

It is vital to note, though, that this was a problem long before we all began to become more aware of it with the furor over COVID. When the first edition of this book was released in Europe in late 2019 as *The Irrational Ape*, coronavirus was not a household term. Even then, the book was replete with myriad stories of how and why we get things wrong, which resonate all the louder thanks to our recent history. Some of the problems we face are age-old: Our psychological quirks make us perennially vulnerable to emotive and incendiary fictions. We remain dangerously pliable to manipulation, a fact that has not gone unnoticed by malicious actors in arenas from politics to medicine. Our self-curation of modern media has led us to deeper divisions and mutual hostility than ever, and we are easily misled by the misrepresentation of everything from logic to statistics.

When the book was released in Europe, the subject was undoubtedly timely. And as the truly staggering events of the last year make clear, the topics explored within have only increased in urgency and importance. The motivation behind this book is to explain why we go so badly askew with our reasoning, and how we can steel ourselves against these frequently damaging missteps. Understanding why we err is critical if we're to circumvent the worst consequences of unclear thinking: Unless we are acutely aware of where we go wrong, we can easily blunder into mistakes that cause us real harm. Critical thinking is the only shield we have against the machinations and ineptitude of charlatans, fools, and demagogues. Our real challenge now and in the future is not accessing information, but rather learning how to critically evaluate the deluge of claims to which we're subjected.

Of course, we are humans—not unfeeling machines. Every one of us is prone to the same faults in our thinking, affected by the same blind spots, imperiled by the same errors. This is not a textbook; it's a collection of incredible stories, delving into the missteps that underpinned them. Stories resonate with us so much more than abstract facts. They echo around our thoughts far after technical musings have atrophied with time. The tales within not only showcase where we can go awry, but also highlight just how bizarre, comical, or tragic the impacts of bad thinking can be. In this, the first American edition, I've expanded and rewritten the text with many new tales to reflect the momentous events of the past year.

Importantly though, none of this should be taken as an admonishment——there is absolutely no shame in being wrong or making mistakes. The only shame is in refusing to correct ourselves when we get things wrong. Perfection in thinking is not the goal——it is enough for us to strive toward being less wrong. To quote my countryman Samuel Beckett, "Ever tried. Ever failed. No matter. Try again. Fail again. Fail better." In this spirit, I hope you enjoy *Good Thinking,* and garner something from it that gives you new tools in the struggle to always be a little less wrong.

David Robert Grimes
November 30, 2020

PROLOGUE

As heroes go, Stanislav Petrov is hardly a household name—it does not leap from our lips, nor does it adorn monuments. Yet every one of us alive probably owes our existence to this obscure Russian.

Why? Well, on September 26, 1983, Petrov was a lieutenant colonel in the Soviet Air Defense Forces. He was serving as the chief officer on duty at Serpukhov-15, a bunker just outside Moscow. This facility was home to OKO, the Soviet missile early warning system—Russia's eye on its enemy. These were fraught times. The Cold War was at its zenith, and deployment of US nuclear-missile systems across Europe had enraged the Kremlin. Tensions between the United States and the Soviet Union had never been higher. Just weeks before, the Soviets had shot down a South Korean civilian flight, killing all 269 passengers on board—including a US congressman.

With President Reagan denouncing the Soviet Union as "an evil empire," relations between the two superpowers had deteriorated to a state of alarming brinkmanship—and whispered in the corridors of power on both sides was the very real prospect of nuclear war. It is difficult to overstate the incredible firepower under these rival nations' command. The first half of the twentieth century had seen physicists uncover the secrets of nuclear fusion, discovering how stars produce their incredible energy. Over subsequent decades both the USA and USSR had spent vast fortunes exploiting this, not for the betterment of humankind but to craft nuclear arsenals capable of obliterating whole cities. With such dreadful firepower, there could be no victors—only survivors.

Against this backdrop, alarms at Serpukhov-15 began their mournful September wail, signaling that five American missiles were inbound. The unthinkable had become reality: Nuclear war was imminent. Stanislav Petrov had long drilled for such an occasion and his instructions were clear: It was his duty to inform his superiors that war had begun. Their response would be inevitable: The Russians too would unleash a volley of nuclear warheads. The Soviet Union would be destroyed, but they would in turn destroy America. In the crossfire every other nation on Earth would be targeted by both superpowers, seeking to nullify any potential advantages for any rival that might survive and vie to rule the ashes.

Petrov was all too aware of this grim future. He also knew that once this news was elevated up the chain of command, the Soviet Union's military commanders would waste no time scrambling to destroy their nemesis in retaliation. Every moment he delayed risked ceding more of an advantage to the American assault, a fact that couldn't have escaped his fellow officers. To them, this was no time for reflection—it was time for decisive action. In this crucible of relentless pressure, Petrov made a different choice. Instead, he called the duty officer and calmly reported OKO as faulty. His colleagues were aghast, but as chief officer his word was final. There was nothing to do now but wait to see whether the lieutenant colonel was correct or whether they would be incinerated.

That we are here is proof that Petrov's instinct was vindicated. His reasoning had been simple and elegant: Were the United States to launch an attack, it would have had to be all-out. They would have had to overwhelm the USSR's missile defenses in the hope of wiping their opponent from the face of Earth. They would have known that Russia would reply in force. If an attack were to come, it would have had to be an almost unfathomable barrage. Yet a paltry five missiles was a far cry from this strategy. Nor had the ground radar picked up any corroborating evidence. Weighing up the probabilities, Petrov had therefore arrived at the conclusion that a malfunction was a much more likely explanation. As it would later transpire, his reasoning was entirely correct—the ominous warheads seen by OKO were nothing more than reflections from low clouds, misinterpreted by the detector.

Petrov's insistence on reasoning before reacting had averted total nuclear annihilation. By all rights, he deserved to be celebrated as a hero the world over. Instead he was reprimanded, ostensibly for failing to document his actions adequately during the crisis. This was an impossible ask, as he recalled years later: "I had a phone in one hand and the intercom in the other, and I don't have a third hand." In reality, Russian military command was embarrassed by the failure of their cutting-edge system and eager to spread the blame. Feeling scapegoated, Petrov eventually suffered a nervous breakdown. He left the military the following year, joining a research institute. Beyond the upper echelons of the Soviet military, no one knew about his actions, nor how close to destruction we had come. It wasn't until 1998 that the world learned of Petrov. Even then, he remained modest, claiming right up until his death in 2017 that he had only been doing his job. Perhaps so—but think of what might have transpired had a less reflective individual been in command.

This was far from the only close call of the Cold War. Two decades before the OKO affair, on October 27, 1962, at the height of the Cuban missile crisis, something even more alarming transpired. While Khrushchev and Kennedy engaged in frantic diplomacy to prevent war, another crisis was simmering deep beneath the surface of the North Atlantic Ocean, unknown to either leader. The Soviet submarine *B-59* had been detected by the US Navy, and in response dived too deep to communicate with the outside world. Pursued by the aircraft carrier USS *Randolph* and 11 destroyers, the *B-59* crew had been unable to contact Moscow for days. No one aboard had any idea if war had begun or how to proceed.

In an attempt to force the submarine to surface for identification, the Americans then began dropping depth charges, which was unsurprisingly interpreted by the Russians as an act of aggression. The three senior officers on board—Captain Valentin Savitsky, political officer Ivan Semonovich Maslennikov, and flotilla commander Vasili Arkhipov—gathered to formulate a response. Cut off from Moscow, B-59 had autonomy to respond to threats and, if required, the authority to deploy the single T-5 nuclear torpedo in the ship's arsenal. This was a nuclear capability of which their American pursuers were entirely unaware as they continued hounding the beleaguered sub.

Aboard B-59, the atmosphere was oppressive. The air conditioning had failed and the already cramped enclosure was like an inescapable sauna, with temperatures climbing above 122°F (50°C). Carbon dioxide had risen to dangerously high levels, and oxygen was low—neither situation conducive to rational decision-making. Drinkable water was in short supply too, and crew members were restricted to a single glass of water a day. With American depth charges constantly rocking B-59, intelligence officer Vadim Orlov later described how each barrage "felt like sitting in a metal barrel with someone hitting it with a sledgehammer." In such hellish conditions, the rattled Captain Savitsky accepted that war had already begun. "There may be a war raging up there and we are trapped here turning somersaults. We are going to hit them hard. We shall die ourselves, sink them all, but not stain the navy's honor," he proclaimed, ordering his crew to target the USS *Randolph* with the 15-kiloton nuclear torpedo.

Maslennikov agreed. Normal protocols dictated that a decision to launch required the approval of the captain and political officer only. But Arkhipov's position as flotilla commander gave him equal rank with Savitsky. For B-59 to use its nuclear weapon, all three would have to consent. With Savitsky and Maslennikov resolved to fight, the decision to strike now rested entirely upon Arkhipov's broad shoulders. Upon his word, the *Randolph* would have been completely vaporized by the nuclear payload, an act that would have triggered a Third World War. Neither the Kremlin nor the White House knew that this momentous decision was being made. In the words of historian Arthur M. Schlesinger Jr., "this was not only the most dangerous moment of the Cold War. It was the most dangerous moment in human history."

The commander was, however, no stranger to pressure. Only the year before, he had served on the *K-19* submarine, when its nuclear reactor coolant system failed. To stave off a nuclear meltdown, Arkhipov and the crew had improvised a secondary coolant system that narrowly averted disaster. In the process, the crew had received incredibly high doses of radiation. Although many succumbed to radiation poisoning, meltdown had been narrowly avoided. This incident was infamous throughout the Soviet navy, and Arkhipov's courage was widely known and deeply respected. Now, aboard the sweltering *B-59*, all eyes fell upon him. Facing his fellow officers, he resolutely vetoed their request to engage. A passionate argument ensued, yet his contention remained that launching the T-5 meant total nuclear war was inevitable. To do so without complete information was the height of madness, he argued; instead, he urged that they surface and reestablish communication with Moscow.

In the end, Arkhipov won his colleagues over. By that stage the White House had become aware of the North Atlantic chase, and gave orders that *B-59* be allowed return to the Soviet Union unmolested. It was not until much later that either Moscow or Washington had any inkling of quite how close to destruction the world had come, and how Arkhipov's level head had prevented Armageddon. Decades later, the director of the National Security Archive, Thomas Blanton, put it succinctly: "A guy called Vasili Arkhipov saved the world."

While Petrov and Arkhipov may never receive the recognition they deserve, humanity owes them a huge debt. Their actions share something too, quite aside from the fact that each of them averted doomsday. In situations where emotions ran high, both men employed impressive critical thinking, and quite literally saved the world. In the face of incredible stress, they marshaled logic, probability, and clear reasoning. And because of that, we are here today. We may never ourselves have to avert a nuclear disaster, but we ought to learn something from these two unsung Russians: that the ability to think critically is absolutely vital.

INTRODUCTION

FROM ABSURDITY TO ATROCITY

China of the 1950s was a country in rapid flux. After a hard-won victory, the Communist Party was determined to transform an agrarian society into a modern communist utopia. To this end, Party Chairman Mao Zedong hatched an audacious plan: the Great Leap Forward. Mao's vision for this rapid industrialization required collectivization of farming and a suite of new policies. It was deemed imperative that vermin be eliminated—the flies that pestered humankind, the mosquitoes that spread malaria, and the rats that propagated plague. This rogues' gallery was rounded off by a perhaps unexpected inclusion: the humble Eurasian tree sparrow. This harmless bird vectored no disease, but it ate grain the farmers had sown. To the authorities, sparrows had political resonance, a parasitic bourgeoisie exploiting the proletariat. With the birds denounced as "public animals of capitalism," the Great Sparrow Campaign of 1958 aimed to exterminate these winged enemies of the revolution.

The *Peking People's Daily* demanded that "all must join battle ... we must persevere with the doggedness of revolutionaries." This call to arms was emphatically answered; Beijing alone mustered a 3-million-strong force. Student rifle teams were trained to shoot sparrows, nests were systematically destroyed, eggs broken, and chicks killed. Others banged pots, with the resultant cacophony preventing the sparrows from landing. Exhausted, the poor creatures fell dead from the sky in droves. Terrified birds flocked to anywhere they could find sanctuary, such as the Polish embassy in Beijing, which refused the mob entry. Any respite was short-lived, as the grounds were surrounded by volunteers beating drums. After two days of constant drumming, the Polish mission had to use shovels to clear the dead sparrows. Within a year, an estimated 1 billion sparrows had been killed, rendering them virtually extinct in China.

But the architects of this destruction had not considered the importance of the simple sparrow. Autopsies revealed that their major food source wasn't grain but insects. Nor was this unforeseen—China's leading ornithologist, Tso-hsin Cheng, had warned that sparrows were vital for pest control. This perceived criticism incurred the wrath of Mao, and Cheng was branded a "reactionary authority," condemned to reeducation and hard labor. The party eventually yielded to reality in 1959, but the damage was done. Sparrows were the only natural predator of locusts and, in their absence, insect populations exploded. Across the country, locusts devastated crop yields unimpeded. This havoc forced China into a spectacular about-face, importing sparrows from the Soviet Union. But crop yields were already irreparably damaged, a situation exacerbated by other disastrous policies of the Great Leap Forward. The ensuing result of this myopia was the Great Chinese Famine between 1959 and 1961, a tragedy that claimed between 15 and 45 million innocent lives.

This staggering loss of life is a stark illustration of the failure of thought, a testament to what can happen when actions are pursued without reflection for what the consequences might be. Mao and his contemporaries were taken in by the politician's syllogism, "Something must be done; this is something; therefore, this must be done." But simply taking action for its own sake is no guarantee that action will be beneficial. As the adage warns, the road to hell is paved with good intentions; poorly considered actions can lead to unintended, dreadful results. The party's overwhelming desire for modernization had blinded them to the dangers and rendered them deaf to the concerns of the scientists who had urged caution. The Great Chinese Famine is an example of what can transpire when critical thought becomes afterthought.

Our ability to reason, reflect, and infer is one of our finest skills and perhaps what best characterizes us as a species. It is conceivably the secret of our success. Our total dominance of this planet is in some respects surprising. As a species, we are not especially imposing—we are furless, bipedal apes, possessing only meager physical prowess. We cannot deftly scale trees like our simian cousins. Nor do our physiques compare favorably to the sleek powerful forms of hunting predators. In our natural state, we are confined to Earth, incapable of flight, unable to survive long in open water—and even less time submerged in it. But our greatest endowment is just about two pounds (1 kg) of fleshy matter with the consistency of gelatin, encased in the protective fortress of our skull. Since humankind took its first tentative steps on the planet, the extraordinary power of our unique brains has been the one feature that has allowed us rise to the apex position, more than compensating for what we lack in tooth and claw.

An intricate dance of chemical and electric signals inside our heads has given rise to all that makes us human. Language, emotion, society, music, science, and art all come from our ability to think and to share those thoughts. This ability to communicate and our limitless capacity for reason have led us to extraordinary feats. Our minds have enabled us to shape the world around us, bending nature to our will. We are—and always have been—driven by curiosity, deep thought, and an irrepressible desire to explore. We possess an insatiable hunger to discover more about the majestic world around us, to better understand our place in the vastness of the universe. We have traversed the deepest oceans, unlocked the secrets of the atom, and even escaped the confines of our planet. The very name of our evolutionary niche reflects these traits—*Homo sapiens;* the thinking man—as much a statement of intent as a description.

But for all the virtues of our minds, faults in our reasoning are pervasive. Despite the impressive hardware with which we are gifted, we frequently make mistakes ranging from trivial to fatal. While these have blighted us throughout history, it is now more vital than ever that we understand where we can err. We have never been more at the mercy of charlatans and fools, from fraudulent health advice to the emergent phenomena of fake news and viral propaganda. These are not new problems, but the scope of the challenge has changed utterly. We live in an era where instantaneous access to the wealth of human knowledge is at our very fingertips. Yet the paradox is that this same freedom allows misunderstanding, misinformation, and falsehoods to perpetuate further and faster than ever before.

But we needn't despair—the same human mind that can make mistakes is also uniquely capable of learning from those mistakes. If we can identify where we err, then we can circumvent the consequences of faulty thinking. If we are to make sound decisions in the face of an overwhelming cacophony of half-truths and outright lies—the equivalent of all those banging pots in the Great Sparrow Campaign—it is imperative that we learn to distinguish how to separate the signal from the noise and be aware of where faulty reasoning might creep up on us. Daunting as this might seem, we have an extraordinary advantage: the ability to think critically. There are many related definitions of this—the *Oxford English Dictionary* defines critical thinking as "the objective analysis and evaluation of an issue in order to form a judgement."

The analytical aspect of this is vitally important. If we can learn how to trace the path of each assertion to its logical terminus, we can derive much more reliable conclusions than instinct or intuition alone would allow. Perhaps more difficult is to subject our own beliefs to the same scrutiny we'd apply to the convictions of others. We must let evidence guide us, and be prepared to jettison incorrect ideas and beliefs, no matter how comforting they might be. The question isn't whether we like the resulting conclusion or whether it fits our preferred view of the world; only whether it flows from the evidence and logic or not.

Such reflection is vital, as our view of the world is inherently skewed. Swedish statistician and physician Hans Rosling surveyed thousands of people worldwide, asking them objective questions about everything from health care to poverty. His repeated finding was that, no matter our level of intelligence or education, we are resoundingly underinformed about the world. We harbor impressions totally incompatible with the data, and these impressions are frequently far more pessimistic than the evidence implies. In Rosling's view, this is due to our tendency to rely on media accounts to form impressions, remarking that "forming your worldview by relying on the media would be like forming your view about me by looking only at a picture of my foot." Media of course entails much more than the traditional triumvirate of television, newspapers, and radio. The majority of us now get our news and information online, overwhelmingly through social media. "Stripped of the gatekeepers and regulations constraining traditional media, this is an environment in which distortions, misconceptions, and outright falsehoods can quickly take root."

Nor are we especially skilled at detecting falsehoods. In 2016, researchers at Stanford tested the ability of middle-school, high-school, and university undergraduates to gauge the credibility of different articles. The results were, to quote the researchers, "bleak" and a "threat to democracy." Across the board, students were easily misled into accepting dubious sources as legitimate, unable to even identify what they needed to look for to assess the legitimacy of the source. The simple fact that a website "looked" polished or a social media account had a lot of followers was enough to dupe even these digital natives. Stanford undergraduates, for example, were directed to articles on same-sex parenting from the American Academy of Pediatrics (a reputable professional body) and the American College of Pediatricians, a recognized homophobic hate group. Depressingly, the students saw the two organizations as equally reputable, failing to look beyond the website or do rudimentary fact-checking.

An estimated 59 percent of articles shared on social media are propagated by people who haven't even read them. Reading an article takes effort, whereas sharing something based on an appealing headline alone garners social kudos without any intellectual exertion. This social component is deeply important; more so than traditional media, online sharing caters to our worst excesses. A 2014 study in *Science* found that learning about immoral acts online triggered far stronger feelings of outrage than when the same acts were reported on television or in a newspaper. Part of the reason is that content producers and platforms are reliant on sharing to generate revenue. Even traditional non-tabloid media—whose revenue streams once depended on trusted reporting—have been forced to embrace the internet as physical sales plummet. And the best predictor of online sharing? Strong emotions. A 2017 PNAS study found that moral-emotive language significantly increased the diffusion of political content across social media. But this comes at the cost of turning us into engines of outrage, implicitly selecting for the most arresting content, regardless of its veracity or social value.

Cathartic and emotionally justified as shared outrage may sometimes be, it's not conducive to finding viable solutions. If anything, it drives us deeper into our tribes; strong feeling might generate more engagement, but this tends to stay within ideological group boundaries rather than transcending them. Nuance is lost, disagreements amplified, and distortions of opposing positions become commonplace. This preaching to the choir gives us a sense of satisfaction, but is ultimately performative if it is not accompanied by meaningful attempts to address the issues. Anger is not a sophisticated emotion; it's a prism that distorts nuanced situations into misleading binaries, and complex characters into heroes or villains. A growing body of evidence suggests that the decline in traditional media has seen an alarming fragmentation of information. By curating our own sources, we can construct any tableau we desire. But collectively we fail to objectively interrogate our information, amplifying that which affirms our prejudices and preexisting beliefs while excluding that which might challenge them. To borrow from Paul Simon, "a man hears what he wants to hear and disregards the rest." The instantaneous nature of modern discourse means we are primed to crave velocity over veracity, reaction over reflection.

The net result of all this should concern us deeply. A massive 2018 study published in *Science* delved into the fractured fabric of modern discourse, analyzing 126,000 contested news stories between 2006 and 2017. Their findings make for sobering reading. By any metric one employs, hoax and rumor completely eclipse truth, and falsehoods consistently dominate the narrative: "Falsehood diffused significantly farther, faster, deeper, and more broadly than the truth in all categories of information, and the effects were more pronounced for false political news than for false news about terrorism, natural disasters, science, urban legends, or financial information." Emotional content was a predictor of how widely shared an item would be, and false narratives were crafted to elicit disgust and fear, and to direct anger.

False narratives foster mistrust, leaving us more polarized than ever before. More than that, though, they're resilient to correction—it takes considerably greater effort to debunk a myth than it takes to craft it in the first instance. This hasn't escaped the notice of propagandists the world over, who have taken advantage of the internet to spread all manner of suspect messages. Russia under Vladimir Putin has been by far the most enthusiastic adopter of this new front—fingerprints of substantial Russian interference have cropped up around the globe, aimed at exerting a destabilizing influence on perceived rival nations by stoking internal tensions and mistrust. One infamous example is the Internet Research Agency outside St. Petersburg, where a small army of trolls is employed to prowl social media, sowing discord and influencing opinion worldwide. A joint report by the United States Intelligence Community found that the 2016 US election was rife with Russian meddling, with subsequent analysis suggesting this concerted propaganda effort

might have been enough to swing the result. Similar telltale signs
of interference cropped up the same year during the UK Brexit
referendum and in the 2017 French presidential race.

Nor do these devious tactics confine themselves to just the
obviously political; the emergence of COVID-19 in 2020 has seen a
surge of deliberate fictions. Such is the extent of the problem that
the World Health Organization deemed the storm of falsehoods an
infodemic—"an overabundance of information, some accurate and
some not, that makes it hard for people to find trustworthy sources
and reliable guidance when they need it." Cynical as it is, much of
the deliberate and inflammatory fictions about the virus stem from
China and Russia. These are propagated, in the words of a European
Commission report, with intention to "aggravate the public health
crisis in western countries, specifically by undermining public trust
in national health care systems."

The depressing truth is that these techniques, cynical as they
are, are incredibly effective. The RAND Corporation describes
this as the "Russian firehose" model of propaganda: high-volume,
multichannel, and unrelenting. While the material lacks any commitment to objective reality or consistency, its rapid and repetitive
nature captures our attention. Things seem more convincing when
they come from multiple sources, pointing to the same conclusions—even if the claims themselves are inconsistent. The principle
isn't always to persuade but to overwhelm us with conflicting narratives until we end up sleepwalking into a state of confused inertia.
The combined effect of all this has disproportionate influence
on what we believe. This is a precarious state—Voltaire famously
warned that "those who can make you believe absurdities can make
you commit atrocities."

The US Office of Strategic Services (OSS) would have agreed
with Voltaire's assessment almost 200 years later. Their psychological profile of Adolf Hitler, commissioned in the midst of the Second
World War, makes for compelling reading:

His primary rules were: never allow the public to cool off; never admit a fault or wrong; never concede that there may be some good in your enemy; never leave room for alternatives; never accept blame; concentrate on one enemy at a time and blame him for everything that goes wrong; people will believe a big lie sooner than a little one; and if you repeat it frequently enough people will sooner or later believe it.

The OSS report doesn't just give a portrait of the most infamous and terrible dictator in history, it captures the blueprint for tyranny itself. Dictatorship can only thrive by subverting our critical faculties, homing in on our biases, and exploiting the glitches in our cognitive mesh. Hitler was a devious and skilled orator who knew intuitively what psychologists refer to as the illusory truth effect— our tendency to believe information to be correct due to repeated exposure. He was certainly not the first to realize this; Napoleon Bonaparte is widely believed to have remarked that "there is only one figure in rhetoric of serious importance, namely, repetition." Research indicates that the simple repetition of falsehood doesn't just bamboozle us on topics over which we're uncertain, it can in some instances even sway us to accept a fiction despite knowing the correct answer.

That our very reality can be so easily eroded is a disconcerting concept, and we are witness to this even now in contemporary politics. All of this is inherently damaging not only to our understanding of the world around us, but to societal cohesion itself. Pervasive falsehoods fracture our trust in society, institutions, and each other. And all too often, devious fictions rush in to fill the void left by suspicion and mistrust. As if to compound all this, we as a species face daunting challenges that demand considered action, from the rapid encroachment of climate change to the resurgence of Cold War geopolitics and the impending catastrophe of antibiotic resistance. Never in human history have our actions had such long-lasting consequences.

For all the sophistication of our minds, we are but reflective animals. We are irrational apes, deeply wedded to questionable conclusions, prone to thoughtless reaction. We have constructed tools of unimaginable destruction and placed them at the whim of volatile tempers. As the great biologist E. O. Wilson suggested, humanity's real problem is that we have "Paleolithic emotions; medieval institutions; and god-like technology."

All of us, of course, harbor some delusions or questionable beliefs. But we might not be aware of quite how drastically these can alter our perceptions. Ideas do not exist in isolation, nor beliefs in a vacuum. All information we encounter forms part of what W. V. Quine called our "web of belief." Our ideas are deeply entangled, and accepting even one dubious belief can mean a spiral of impacts on other concepts. Taking the example of the debunked claims that vaccines cause autism, philosopher Alan Jay Levinovitz elaborates:

> In order to add "vaccines cause autism" to your web of belief, you must weaken confidence in [scientific authorities], and increase the force of other higher order beliefs so they can supply adequate alternative justification. To those who follow the debate over vaccines, these higher order justificatory beliefs are all too familiar: natural is better than unnatural; scientists are in the pockets of Big Pharma; mainstream media can't be trusted; you are the best judge of what's good for your body.

Something similar is seen with conspiracy theorists; belief in one conspiracy theory is strongly correlated with belief in others. Once someone yields to conspiratorial ideation, they begin to see sinister machinations everywhere.

All this leaves us polarized and divided. Democracy itself is fragile—we share but one world, and if we cannot even agree on basic facts, how can we hope to find pragmatic solutions to the problems confronting us? The solution is to adopt the critical thinking central to the scientific method, where ideas are advanced and rigorously tested. Those that withstand critical examination are provisionally accepted, while those that do not are discarded, no matter how elegant they may be. There is nothing inherently scientific about this approach—in essence, it's simply a scientific context for a much more general stratagem where we test our ideas rather than accept them blindly. Conversely, this means this critical approach isn't limited to scientific questions—analytical thinking can be applied to problems across all spheres, from making choices about our well-being or deciding what insurance to buy to averting global disaster. Learning to think like scientists unlocks the tools we need to assess the onslaught of assertions we encounter, untangling whether they're reasonable or suspect. And crucially, it allows us to recognize dubious arguments and misleading techniques.

This not only enables us to make better decisions, it is fundamental to our freedom; critical thought is anathema to demagogues. In a 1995 essay, the great Italian novelist and philosopher Umberto Eco enumerated 14 properties common to all fascist ideologies. His observations were drawn from historical authoritarian regimes, but it is disturbing to note the dark renaissance of many of these traits in modern populist political movements. Chief among these is an odious strain of anti-intellectualism and irrationalism, which seeks to denigrate critical thought. To fascist-like movements, Eco noted:

> Thinking is a form of emasculation. Therefore culture is suspect insofar as it is identified with critical attitudes. Distrust of the intellectual world has always been a symptom of Ur-Fascism, from Goering's alleged statement ("When I hear talk of culture I reach for my gun") to the frequent use of such expressions as "degenerate intellectuals," "eggheads," "effete snobs," "universities are a nest of reds."

That such movements aim to stifle critical thought and denigrate those who encourage it is unsurprising. A society that is willing to ask for evidence and to challenge inaccurate claims, and that is aware of duplicitous tactics, is immune to the arsenal of eager tyrants. This kind of analytical thinking isn't entirely natural to us–it demands that we be reflective rather than reactive, and value veracity over velocity. While not intuitive, it can be learned.

One might assume that rationality is a by-product of intelligence, but there is little correlation between intelligence and rationality. Those with high IQs are as likely to suffer from dysrationalia (the inability to think and behave rationally despite possessing the mental faculties to do so) as those of lower intelligence. Unlike IQ, however, rationality can be readily improved. An intriguing 2015 paper assessed the susceptibility of subjects to common decision-making biases. Afterwards, some subjects were shown an explanatory video on their logical mistakes or asked to play an interactive game designed to decrease bias. Confronted with similar problems months later, those with this training were far less likely to repeat their errors–and far more likely to spot questionable claims.

As a scientist, I have been extraordinarily privileged to have received years of training in analytical thinking. Even now, I still learn new things, and correct old errors. As a science communicator, I've had the additional pleasure of talking with a wide variety of people on their understanding of science and medicine, garnering some insight into their concerns, misgivings, and confusions. I've spent much of my time over the past few years attempting to bring clarity to issues contentious in the public mind, from cancer myths to climate change to vaccination and genetic modification. I've witnessed the darker side of tortuous logic and irrationality: conspiracy theories, misguided crusades, and even needless deaths. And in all of this, there are lessons we can learn that might make us just a little more astute.

My aim with this book is to illuminate the major reasons why we err, and to explore how each of us can employ analytical thinking and the scientific method to improve not only our own lives, but our world itself. It's probably foolishly ambitious to hope to capture all this in a single work, of course, but I hope that this contribution showcases the major issues and ways of thinking that consistently lead us astray. It isn't my intention to write a textbook—stories have deeper resonance for us than facts alone, and so every topic we'll explore is illustrated through strange, true stories from across the world and history, from the comical to the catastrophic.

Accordingly, the book is organized into six major sections, unified by common themes. Part 1, Without Reason, explores our ability to reason. This is perhaps one of humankind's greatest assets, and yet an illusion of logic can drive us to terrible consequence. These chapters focus on the vital importance of logic, and how subtle errors can steer us toward disaster. Part 2, The Pure and Simple Truth?, concerns the perpetual maelstrom of arguments, discussions, and debates to which we're subjected, exploring how rhetoric skews our ability to think clearly, leaving us vulnerable to demagogues and charlatans.

Part 3, Trapdoors of the Mind, reveals how we are unreliable narrators of our own lives. Our very thoughts, emotions, memory, and senses are more malleable than we might know, and here we examine the hidden biases, psychological quirks, and flawed perceptions that push us to faulty conclusions. Part 4, Lies, Damned Lies, and Statistics, delves into the ubiquity of statistics and numbers in the modern world—and just how the true meaning of the figures we encounter is frequently misunderstood or distorted, our collective innumeracy frequently exploited by the duplicitous.

How and where we acquire our information itself plays a huge role in shaping our perceptions. Media has a greater impact on our understanding than we comprehend; in part 5, News of the World, we'll see how what we consume shapes our perceptions, from television to social media—and just how easily we are misled by our own sources. Finally, part 6, The Candle in the Dark, focuses on critical thinking and the scientific method—and how we can use these tools to enlighten our world. These chapters elucidate the fine line between science and pseudoscience, the extraordinary power of skepticism, and how a modicum of critical thought improves our decisions and might yet save the world.

I would hate to give the impression that scientists are flawless—nothing could be further from the truth. We are human, prone to the same errors as anyone else. That we will make mistakes is inevitable, but we can learn from them. Analytical thinking and the scientific method itself are not the preserve of science—they are the property of all of us. Scientists shouldn't be jealous gods atop Olympus but heirs to Prometheus, eager to share fire. We live in an age where the ability to differentiate signal from noise has never been more urgent nor more difficult—an era where myths and manipulations threaten to strangle truth—and so I truly believe that it has never been more important that we embrace analytical thinking whether we're artists or accountants, police officers or politicians, doctors or designers. We'll begin with something fundamental to being human: reason itself.

PART 1

Without Reason

Formal Fallacies and How to Defeat Them

"He, who will not reason, is a bigot; he,
who cannot, is a fool; and he, who dares not,
is a slave."

—WILLIAM DRUMMOND
OF LOGIEALMOND

AN INDECENT PROPOSITION

Errors in Logic, Murderous Popes,
and 9/11 Conspiracies

Strange as it sounds, the medieval papacy was a hive of political machinations as devious as an episode of *Game of Thrones*. But even by the bizarre standards of early Vatican intrigue, few episodes in the history of the Catholic Church are quite as strange as the dramatic events of January 897. The setting was the courtroom in the magnificent Archbasilica of St. John Lateran, where the newly anointed Pope Stephen VI thundered accusations of perjury, corruption, and sin at his predecessor, Pope Formosus. Yet, despite the animated tirade, Formosus reacted with stony silence to the litany of abuses leveled against him. This silence was perhaps unsurprising; Formosus had in fact been dead a full eight months before the trial even began.

Even so, the disinterred Formosus sat propped up, garbed in papal vestments, a perplexed deacon appointed to speak for him. To the shock of absolutely no one, Formosus (whose papal name, somewhat unfortunately, translates as "handsome"–unlikely an apt moniker that long post-mortem) continued his defiant silence. By the rationale of his papal accusers, this silence was damning evidence of guilt. After all, Stephen declared, an innocent man would defend himself. As Formosus did no such thing, he was surely guilty. And so, guilty Formosus was found–Stephen wasted no time in condemning the thoroughly deceased pope, ordering three of the fingers on his right hand to be severed so that he might not perform any blessings, on the off-chance Formosus might add reanimation to his list of achievements.

Formosus's mutilated corpse was flung into the raging Tiber, retrieved by monks and briefly worshipped as miraculous by Roman citizens. The macabre spectacle became known as the

Cadaver Synod or the *Synodus Horrenda*, turning public opinion against Stephen.* Of course, Stephen wasn't a complete idiot—the true motivation of the trial had been nakedly political. Skewed logic was merely used to justify the whole sordid affair, giving the appearance of reason to an episode devoid of any justice. Not that it helped Stephen in the long run; before summer 897 was over, he himself was imprisoned and strangled to death in his cell. The church later quietly disregarded the *damnatio memoriae* against Formosus as based more on politics than piety, wisely letting the whole ugly incident fade quietly with the fullness of time. But there is a fascinating lesson underpinning it all—how we can be misled by the illusion of reason.

Our capacity to reason is the clearest hallmark of being human. We are reflective animals, blessed with metacognition to be aware of that fact. Each one of us wrestles with concepts both abstract and tangible, learning from the past and preempting the future. And underpinning it all is our ability to reason, a spark that illuminates even the darkest reaches. But for all the impressive feats of which our brain is capable, it isn't an infallible machine and we frequently make mistakes both obvious and subtle. Psychologists Richard E. Nisbett and Lee Ross remarked of this glaring contradiction that "one of philosophy's oldest paradoxes is the apparent contradiction between the great triumphs and the dramatic failures of the human mind. The same organism that routinely solves inferential problems too subtle and complex for the mightiest computers often makes errors in the simplest of judgments about everyday events."

Possessing a powerful brain is not enough. We need also to train it sufficiently to handle more obtuse and complex situations.

---

* Formosus was eventually rehabilitated and re-interred in pontifical vestments, but this was not the end of his tribulations. Years later, the ruthless, lecherous Pope Sergius III overturned the pardons. Some sources state he even had the dead Formosus decapitated, just to be sure. The truth of this is hard to verify, but even by the high bar for viciousness set by some medieval popes, Sergius was especially notorious, described memorably by one contemporary as "a wretch, worthy of the rope and of fire."

Drawing a loose analogy with computers, even with hardware to the highest specification, a machine cannot perform without the requisite software. Our brain's architecture and complexity are second to none, but reasoning goes beyond the intuitive and needs to be learned. Defective reasoning is a gateway to utterly wrong conclusions. "Garbage in, garbage out" is a mantra of computer scientists and hardly a new complaint. Charles Babbage, credited as the father of computing, lamented in the mid-1800s: "On two occasions I have been asked, 'Pray, Mr Babbage, if you put into the machine wrong figures, will the right answers come out?' . . . I am not able rightly to apprehend the kind of confusion of ideas that could provoke such a question."

Humans, of course, are not computers, but something else entirely. While we are capable of incredibly deep thought, we also rely on instinctive techniques to make rapid decisions. For instance, we might gauge whether something is a threat based on its similarity to known threats. Such rules of thumb are known as heuristics and are hardwired into us. These short cuts are not always optimal, or even correct, but are regularly "good enough" for most situations and don't use up vast amounts of relatively expensive cognition. Most importantly, they happen so instinctively that we're rarely even aware of the thought processes leading us to certain conclusions. This impulse has served us well, keeping us alive through millennia of prehistory, where rapid decisions were often a matter of life or death.

The problem, however, is that most of the important decisions that we face today require more nuanced thought. Heuristics, while useful, are often inherently unsuitable for the challenges and questions we face. Whether the question concerns geopolitics or health care, we cannot rely on unconscious instinct to guide our judgments, and a knee-jerk approach in these situations is a sure route to disaster. Most issues we face today as a species are not cleanly black and white with straightforward solutions. Rather, they exist on a spectrum of varying shades of gray, with unavoidable trade-offs. For the most pressing problems we face, there's rarely an obvious optimal solution to be found and our decisions require reflection and revision in the light of new information.

Luckily, we have more than reflex and gut feeling at our disposal—we can reason analytically, marshaling information, logic, and imagination to arrive at conclusions. On a small scale, we do this all the time—we make decisions, we choose paths, we plan futures. But while we might pride ourselves on our logic and rationality, we are not immune to error. Missteps in our thinking have long plagued us, and flaws in our logic can be downright difficult to untangle. To compound this, there is ample evidence that the illusion of logic is frequently enough to lull us into misconception—even if an argument is fatally undermined by some structural slip. The costs of this are manifold in every human sphere from politics to medicine and can cost us dearly, leading to persecution, suffering, and damage both to ourselves and to the world in which we live.

These are far from mere academic concerns; while our wonderful minds have steered us toward who we are today, we remain afflicted by the vagaries of poor reasoning. Identifying where we fail is vital to correct this. The challenges we face today are not trivial—we wrestle constantly with complex questions, perpetually assessing the risks and benefits of everything from medical treatment to government policy. As a collective, we're confronted by monumental existential questions too, from the looming specter of climate change to epidemics and global strife. Our ability to reason is the only chance we have for finding pragmatic constructive solutions to these broadsides, and if we are to address these problems and more besides, we cannot afford indulgence in half-cocked thinking. But what precisely differentiates solid reasoning from a dubious imitation?

This question has captivated inquisitive minds for centuries—early Greek philosophers dedicated huge amounts of time to exploring the structure of logic. Their discoveries remain the very foundation of mathematical logic. This fundamental area has extreme practical application as well as theoretical elegance, underpinning everything from search engines to space flight, pizza delivery to emergency services. The rigors of logic are not just a niche area for scholars and engineers; it is the very basis of the rhetorical arguments that we encounter every day and the tools we use to reach conclusions on every imaginable issue.

For our purposes, we'll define an argument as a sequence of reasoning steps leading to a conclusion. When the structure of our logic is inherently flawed, we're dealing with a class of reasoning error known as the *formal fallacies*. A full treatment would require us to delve into abstract mathematics, but for our purposes we need only concern ourselves with some essential ideas. For an argument to be sound, it needs to have (a) a valid structure and (b) premises that are correct. Validity might be thought of as the structure or skeleton of the argument. A classic example concerns Socrates, widely considered the father of Western philosophy:

Premise 1:   All men are mortal.
Premise 2:   Socrates is a man.
Conclusion: Thus, Socrates is mortal.

This is an example of *deductive reasoning*, where conclusions flow directly from the premises.* Curiously, we have no record of Socrates' writings, instead deriving our understanding from his contemporaries, Xenophon and Plato. How much these accounts reflect his philosophy or whether they describe a man or idealized figure are matters of some contention, and the air of mystery around the man himself is dubbed the "Socratic problem." All we know for certain is that he was put to death by the state of Athens in 399 BCE, poisoned by hemlock. Beyond this, the historical record

---

* There are other types of reasoning too, most importantly inductive reasoning, where premises are given to provide strong evidence rather than absolute proof of the conclusion. In this case, statements are probabilistic rather than certain. We'll mainly concern ourselves with deductive logic, but the points addressed still apply.

is murky. But execution notwithstanding, the argument shows that the eventual death of the great philosopher was inevitable. Crucially, for an argument to be valid, the only condition is that the logical structure is correct, with the premises leading to the conclusion. Let's consider some nonsensical premises:

Premise 1:   Greek philosophers are time-traveling killer robots.
Premise 2:   Socrates is a Greek philosopher.
Conclusion: Thus, Socrates is a time-traveling killer robot.

While outlandish, the logic is valid; accepting the premises means the conclusion follows. Clearly, valid logical syntax alone isn't enough; for a deductive argument to be sound, the logic must be valid and the premises must be true. With these straightforward examples, it's tempting to assume that gauging soundness is simple. Alas, this isn't always the case—as with all things, the devil resides in the details. Formal fallacies are rudimentary errors in the logical structure of an argument that render the argument invalid. Some can be surprisingly opaque, embedded in cunning demagogic oratory. Let's return to the scheming Pope Stephen's argument against his deceased predecessor:

Premise 1:   An innocent man would defend himself.
Premise 2:   Formosus did not defend himself.
Conclusion: Thus, Formosus is guilty.

The conclusion here is inferred from a statement when there are no grounds to do so. There are myriad reasons an innocent person might not defend themselves. Perhaps they're protecting someone or refusing to recognize a corrupt court. Perhaps they're simply exceptionally dead, as was the case with Formosus. This logical fallacy is *denying the antecedent*, or the inverse error. Just because X implies Y ("an innocent man would defend himself"), it is mistaken to assume the absence of X implies the absence of Y ("Formosus did not defend himself, thus he is guilty"). Despite a superficial logical veneer, it is intrinsically flawed. Greek scholars demonstrated the perils of the inverse error in antiquity, but that hasn't stopped it from being dubiously employed in subsequent centuries by those who should know better, as Pope Stephen exemplified.

The problem with logical fallacies like this is that they often give rise to sensible-looking conclusions, masking more serious issues. These can require some reflection to detect. For instance, one can invert cause and effect—if we're told X implies Y, then it might seem reasonable to presume this flows both ways, with Y implying X. Revisiting Socrates again, this extrapolation would be:

Premise 1:   All men are mortal.
Premise 2:   Socrates was mortal.
Conclusion: Thus, Socrates was a man.

Superficially at least, this appears fine—the conclusion passes a simple sanity check, and the premises appear reasonable. But while the conclusion is true, the argument is invalid—we have no reason at all to assume simply because X implies Y, that Y implies X. Such a logical blunder is known as *affirming the consequent* or the converse error. It's surprisingly common, because it often yields ostensibly correct conclusions from a less-than-watertight logical structure. But the "hits" of this reasoning are simply blind chance. The structure of the argument is always invalid, even if it leads to a seemingly acceptable conclusion; replacing "men" in the above with "dogs" would have equally correct premises, but lead to a false conclusion:

Premise 1:   All dogs are mortal.
Premise 2:   Socrates was mortal.
Conclusion: Thus, Socrates was a dog.

Or, taking a more tangible example:

Premise 1:   Paris is in Europe
Premise 2:   I am in Europe.
Conclusion: Thus, I am in Paris.

While this might be true for the 2.21 million residents of Paris, it's clearly false for the vast majority of the 500 million people in Europe. Affirming the consequent here leads to the conclusion that those in Dublin, London, Berlin, Brussels, or multitudinous other places are inside Paris, presumably causing astronomical delays on the Métro and formidable queues for the Eiffel Tower. That this yields the right answer for Parisians is a mere fluke. However, because it can produce misleading hits, it is often employed in arguments, despite flimsy rooting.

The converse error is easy to spot in the examples so far. But if it's employed subtly, even the relatively astute can fall victim to a disguised version. Advertisers rely heavily on an implicit version when hawking luxury items from perfumes to sports cars. Ads typically show successful, attractive people coveting some item, the implication being that desiring that item makes one a successful, attractive person. The logic of such scenarios is that purchasing the product in question ultimately makes a person desirable sexually or socially. Yet, as anyone who has ever seen a rotund middle-aged man in a sports car will attest, this conclusion does not follow.

Appeals to vanity aside, converse errors lend the illusion of justification to darker arguments. On September 11, 2001, four passenger planes were hijacked in the United States by Islamic extremists in a coordinated attack. American Airlines Flight 11 struck the north tower of New York's World Trade Center between the 93rd and 99th floors at 490 mph (790 km/h). Minutes later, United Airlines Flight 175 struck the south tower at a speed of 590 mph (950 km/h) between the 77th and 85th floors. The violence of the impact draped the towers in thick black smoke, consuming them in raging flames, compromising the structures far beyond their limits of endurance.

By 10:30 AM, both towers had succumbed to catastrophic failure, crumbling before a dumbstruck world.

Across the country, the hijackers of American Airlines Flight 77 careered the passenger jet into the Pentagon. In an act of extreme bravery, passengers on United Airlines 93 rushed their hijackers, sacrificing their lives to bring down the plane before it reached its intended target in the political heart of Washington. As the chaos receded over the smouldering ruins, 2,996 people lay dead in the worst terrorist attack ever on American soil. The world reeled at the sheer audacity of the attack at the heart of the world's most powerful nation, permanently etching the image of the mighty Twin Towers coming undone upon our cultural consciousness.

But before the smoke had even settled, allegations of conspiracy were already surfacing. In the aftermath of the atrocity, the absence of easy answers left a void that conspiracy theorists eagerly filled. Dark conjecture grew in the telling, and an elaborate and all-encompassing narrative emerged. Many asserted that burning jet fuel simply would not have been hot enough to melt steel beams. Others insisted the towers were felled in a controlled explosion. The identity of the "true" perpetrators varied with the prejudices of the believer—some asserted that the attack was simply allowed to happen for political currency. Others claimed it was a false flag operation by the US government or the work of Mossad, while others insisted the entire event was an orchestrated ruse, proclaiming the planes were disguised missiles or even holograph-ically projected mirages to fool eyewitnesses on the ground and millions at home.

What began as fringe views held an undeniable allure. In the wake of 9/11, internet conspiracy sites flourished. Just a year after the attacks, marchers in San Francisco decreed angrily that President George W. Bush was behind everything. YouTube uploads asserting all manner of conspiracies were eagerly consumed. One such doc-umentary, *Loose Change,* racked up millions of views. Its popularity transcended digital confines, prompting *Vanity Fair* to declare it the world's first "internet blockbuster." While the kaleidoscope of theories about what really transpired were often contradictory or thoroughly outlandish, they were united by a common belief: The official account could not be trusted. From the ashes of downtown Manhattan, the 9/11 "truther" movement slithered into public consciousness.

That these ideas found a ready audience is understandable. In a paradoxical way they were darkly reassuring, making sense out of carnage that was otherwise impossible to comprehend. If 9/11 was the flame that ignited such ideas, the 2003 invasion of Iraq was gasoline. Flimsy attempts by the Bush administration to link the attack with Saddam Hussein's regime rang insultingly hollow, as no evidence linked the Iraqi dictator and al-Qaeda. Claims that Saddam had weapons of mass destruction transpired to be false. Invading Iraq was profoundly unpopular, with Canada, France, Germany, and Russia opposing war. On February 15, 2003, anti-war protests were held in over 600 cities around the world, attracting between 10 and 15 million people—the largest protest in history. Disingenuous rationalizations by the Bush administration were grist for the mill for conspiracy advocates.

From that sea of anger, 9/11 myths underwent dramatic amplification. In 2003 I was 17, on the cusp of going to college and, like so many others, I protested against the war that was to ensue. Starting college that autumn, I remember vividly a fellow student who held an audience rapt, joining the dots between all manner of events. In his telling, the towers came down in a controlled explosion, a pretext to the invasion of Iraq. Osama bin Laden was a US agent, Saddam Hussein an innocent scapegoat under whom the Iraqi people thrived but whose oil America needed. This student emissary was in no way unique—such narratives played out verbatim to receptive audiences the world over. It seemed so appealingly clean, explanatory, and reassuring. But for all these attributes, such stories were and remain utter nonsense, readily disassembled by even cursory familiarity with the evidence.

To take one persistent canard, it is true that jet fuel cannot melt steel beams. It is essentially kerosene, burning at approximately 1,500°F (815°C), whereas steel's melting point is around 2,750°F (1,510°C). Yet, while 9/11 truthers clutch at this factlet with religious fervor, it simply highlights a profound misunderstanding of basic mechanics: Steel rapidly loses its tensile strength with temperature. At 1,094°F (590°C), it diminishes to 50 percent of normal strength. At the temperatures in the Twin Towers, it would have decreased to roughly 10 percent of normal. In this hellish crucible, the structure was simply too weakened to endure. This, coupled with the massive structural damage, was the catalyst that let floor collapse upon adjacent floor, an effect known as "pancaking," the destruction multiplying with each level consumed. Steel didn't have to melt to cause the tower's demise—it merely had to fail, a finding constantly reiterated by engineers and professional bodies.

The sequential collapse expelled huge volumes of smoke and air, shattering windows along its descent. As flaming kerosene traipsed down the stairs and shafts, pockets of flame were forcibly ejected over the Manhattan skyline, leading to feverish speculation that a "controlled explosion" had taken the towers down. However, controlled demolitions are undertaken from the ground up, not vice versa. In any case, such a scenario would have required tons of explosives to be somehow smuggled into the building undetected.

Viewed through a critical lens, the pillars of faith the 9/11 truther movement rest upon crumble to dust. Comprehensive investigations into the disaster by numerous agencies and outlets—such as the Federal Emergency Management Agency, National Institute of Standards and Technology, and *Popular Mechanics*, among others— have debunked almost every claim made by conspiracy theorists. The 9/11 Commission found that Mohamed Atta had led the attacks, and all hijackers were members of bin Laden's al-Qaeda. They also concluded that Saddam Hussein and Iraq had no role in 9/11, an embarrassment for the politicians who had advocated the nonexistent link as a pretext for invasion.

I may have been more susceptible to stories of controlled explosions, but my father was a structural engineer, patient enough to explain progressive collapse to me. Had I not grown up in Saudi Arabia (where 15 of the 19 hijackers were born) and witnessed the fundamentalist horror of Wahhabism first-hand, maybe I would have doubted that such theological hatred was even possible. Without some familiarity with Iraq, maybe I could have envisioned Saddam as a benign patsy, unaware of his brutality.

I was fortunate to have this context, but what is surprising is how resilient the movement remains to the multitudinous reports and evidence that completely undermine the truthers' position. The truther movement remains strong, immune from the intrusion of abundant evidence undermining its claims. At the time of writing, approximately 15 percent of the American population is convinced 9/11 was an "inside job," while half of Americans believe successive administrations have covered up the full extent of what happened. Even now, years after the attacks, can such a position be deemed tenable? Liberal application of the converse fallacy explains a great deal of this—in the dark underbelly of conspiracy theories, it functions as a universal deus ex machina for hammering nonsense into narrative. While the array of theories proposed by 9/11 truthers have been comprehensively debunked, they persist despite all evidence against them, with truthers resolutely justifying their conviction by a version of the converse fallacy:

Premise 1:   If there's a cover-up, official reports will
               undermine it.
Premise 2:   These reports debunk our claims.
Conclusion: Thus, there's a cover-up.

This logical contortion renders the glaring absence of evidence for such claims as a bizarre supporting argument. It does not seem to matter how many respected and impartial agencies and examiners debunk truther claims—the same faulty logic is employed to disregard them. Indeed, a quick Google search provides literally thousands of sites dismissing "official accounts" of 9/11 with precisely this skewed reasoning. It seems 9/11 "truthers" employ their name without a trace of self-awareness. It's not just 9/11, of course—any paranoid worldview can be superficially justified provided one throws valid argument to the wind and embraces the converse error wholeheartedly. As we shall see throughout this book, it underpins every color and stripe of conspiracy theory.* The logic employed for such intrinsically hollow arguments gives a veneer of superficial intellect to an emotive or ideological argument. Despite these being completely bereft of substance, they can be used to counteract a fact-based argument and are frequently employed for this purpose.

Slaying these myths is a Sisyphean task; new ones arise hydra-like to take the place of the fallen one. As sociologist Ted Goertzel observed: "When an alleged fact is debunked, the conspiracy meme often just replaces it with another fact." The converse error is a shield against the imposition of reality, a totem to preserve belief, no matter how strongly evidence weighs against it. Enduring beliefs in grand scientific conspiracies are an interesting case in point—many believe that the pharmaceutical industry covers up cures for cancer, for example, or that climate change is a hoax perpetuated by scientists; 7 percent of Americans believe the moon landings were faked, and many more suspect vaccination is some sinister government ploy. In these narratives, the common thread is that scientists are complicit in mass deception. Anyone who's spent any time around scientists will no doubt find this amusing, as trying to get scientists to agree is often vaguely akin to herding cats.

I've witnessed beliefs like these many times in outreach work. They materialize with clockwork precision on subjects where public perception is off-kilter with scientific consensus. When I write on

---

* This is not solely a logical fault. Research has consistently shown that conspiracy theories are a staple of both left and right fringe groups, deeply connected to the ideology of the believers—psychological aspects of which we'll explore in subsequent chapters.

topics like vaccination, nuclear power, water fluoridation, cancer, or climate science, a common strategy from fringe elements is to employ the "shill" gambit, insisting I must be a covert agent paid for by industry. This is nonsense, a mere reiteration of the converse fallacy—"a shill would say this; thus, the author is a shill"—deployed so accusers can dismiss information contradicting their position rather than accept they might be mistaken. I've long been fascinated by how pervasive such conspiratorial views are and how they interfere with public understanding of science. This interest led me to write a scientific paper in 2016 on the viability of conspiratorial beliefs, attempting to gauge whether such mass complicity by the world's scientists would even be possible: Could NASA fake the moon landings, or climate scientists perpetuate a global warming hoax? Constructing a simple mathematical model, the inescapable conclusion was that—even if all conspirators were skilled secret-keepers—large conspiracies were incredibly unlikely to endure for any appreciable time frame.

This wasn't a surprising result—while conspiracies undoubtedly occur, keeping large ones secret for long is nigh on impossible. As far back as 1517, Machiavelli advised against them, observing that "many [conspiracies] have been revealed and crushed in their very beginning, and that if one has been kept secret among many men for a long time, it is held to be a miraculous thing." Benjamin Franklin writing two centuries later was even more succinct: "Three may keep a secret, if two of them are dead."

In our interconnected age, it's even more difficult to keep things under wraps. Still, my conclusions jarred with the central tenet of conspiratorial narratives. Within hours of that paper's publication, I was inundated with emails, blogs, and videos, bellowing that my suggestion that there was no overarching scientific conspiracy "proved" I was part of it—a beautiful example of the converse fallacy in action. My experience isn't unique—*argumentum ad conspiratio* (argument to conspiracy) is the default accusation leveled by conspiracy theorists when confronted by those who counter their assertions. One study we conducted in 2020 found that over 70 percent of medical science communicators had endured this precise denigration. Such accusations negate conflicting information without actually bothering to engage with it on any deep level, stemming the cognitive dissonance that contradictions might invite. This is doubly a shame because, as we shall see, contradictions themselves tell us an awful lot about our reality.

2

STRIPPED TO THE ABSURD

The Fallacy of the Undistributed Middle,
5G Panic, and Online Shaming

Imagine being told that steel is lighter than air. You'd object, surely—were that true, steel would be ethereal enough to hover, scattering like dandelion seeds in the wind. Without performing a single measurement, we know this can't be. Our cars don't have to be anchored, nor do battleships behave like balloons. If we accepted the claim, it would lead to untenable contradictions with what we observe. The resulting absurdity means we confidently reject it. This is the essence of reductio ad absurdum (reduction to the absurd), where premises are disproven because they give rise to insurmountable contradiction. In this respect, contradictions are supremely useful, a warning sign that we've erred in our assumptions or reasoning. The great mathematician G. H. Hardy described them as "a far finer gambit than any chess gambit: a chess player may offer the sacrifice of a pawn or even a piece, but a mathematician offers the game."*

The mathematical form has a curious origin, stemming from perhaps one of the most contradictory characters in history—Pythagoras of Samos. More than 2,500 years after his death, his name lives on in the triangular theorem bearing his name.† As

---

* Hardy once boasted that his work had no practical application, of which he was unreasonably proud. The joke is on Hardy, whose work on number theory is central to the cryptography we in the Information Age are dependent on, wonderfully explored in Simon Singh's *The Code Book*.

† Stigler's Law of Eponymy, articulated by statistics professor Stephen Stigler, proclaims that "No scientific discovery is named after its discoverer." Pythagoras's theorem is a prime example, known to ancient Babylonians and Egyptians. Pleasingly, Stigler attributed his law to sociologist Robert K. Merton, ensuring consistency. Mathematics has an undue number of misattributed theorems. Many of these were documented by historian Carl Boyer, prompting mathematician Hubert Kennedy to establish Boyer's Law—"Mathematical formulas and theorems are usually not named after their original discoverers." Kennedy wryly observed that this was "a rare instance of a law whose statement confirms its own validity"—a statement Greek philosophers would no doubt lose sleep over.

well known as his moniker is, the historic Pythagoras was a complex and strange individual, as much mystic as mathematician,
endowed with both curious spiritual doctrine and impressive
ego. More reminiscent of L. Ron Hubbard than G. H. Hardy,
he founded an eponymous religious sect—the Pythagoreans.
The fine detail of their beliefs has inevitably eroded with the
years, leaving only fragments of their doctrine. They were keen
believers in metempsychosis, a Greek version of reincarnation.
According to Xenophanes, Pythagoras was startled by a dog's
bark, which he interpreted as a deceased friend reborn with
canine physiology. Followers of the philosopher-mathematician
abstained from meat and fish, rendering them among the
first documented vegetarians. For some unfathomable reason,
Pythagoras was singularly averse to beans, his acolytes strongly
prohibited from consuming them. Precise reasons for this are
lost in the mists of time, but it is believed that the beans held a
sacred connection to life. This has been extrapolated to claims
that Pythagoras believed humans lost part of their soul when
passing gas.

In Samos, Pythagoras dwelt in a secret cave, and prominent
citizens consulted him on matters of public concern in a school
he dubbed the "semicircle." He spent time in Egypt, influenced by
the symbolism and mystery of their high priests. He established
his sect in the Greek colony of Croton, where initiates were
sworn to secrecy, bound to communal living. Progressively for
the time, women were admitted. Symbolism was of paramount
importance and sacred icons were kept inside the commune.
Strict penalties awaited any devotee foolhardy enough to reveal
them to outsiders, and edicts from the master were often bizarre,
seemingly born of a whim. Followers were commanded never to
urinate facing the sun, nor to pass an ass lying in the street. Still,
Pythagoras's influence is lasting, as Bertrand Russell expounds in
A History of Western Philosophy:

Pythagoras is one of the most interesting and puzzling men in history. . . . He may be described, briefly, as a combination of Einstein and Mrs. Eddy.* He founded a religion, of which the main tenets were the transmigration of souls and the sinfulness of eating beans. His religion was embodied in a religious order, which, here and there, acquired control of the State and established a rule of the saints. But the unregenerate hankered after beans, and sooner or later rebelled.

Unorthodox beliefs aside, the unifying philosophy was the imbuing of mathematical identities with religious significance. To Pythagoreans, numbers exuded divinity, and relationships between them held the secrets of the cosmos. The parallels with religion are not overstated; after discovering a proof for the 47th proposition of Euclid, the Pythagoreans ritually sacrificed an ox. They searched for esoteric meaning in the harmony of numbers and, of all their beliefs, the mystical ratio was valued above all else. The Pythagoreans believed that all numbers could be expressed as a special ratio, a unique fraction with intrinsic mystical properties. For example, the number 1.5 would be reduced to its essential ratio of 3/2, or 1.85 to 37/20. The same logic applied to whole numbers, so 5 would be reduced to the elemental fraction of 5/1.

Numbers that can be expressed as simple fractions like these are known as rational numbers. To the Pythagoreans, it was an article of faith that all numbers were of such a form, and rationality was the rock their spiritual philosophy was anchored upon. Nature itself seemingly confirmed this; Pythagoras and his followers were deeply interested in music, observing that harmony arose when a vibrating string was shortened into neat fractions. You can demonstrate this with a correctly tuned guitar—pluck an open string and let it sound. Now, fret the string at exactly half its length, at the 12th fret marker. The note will be an octave higher than the open string, with double

* Mary Baker Eddy was the founder of the Church of Christ, Scientist.

the frequency. Fretting an electric guitar at the 24th fret halves the vibrating length again, yielding a resultant note two octaves above the open string. These metaphysical insights into tuning and harmony provided further evidence for the divinity of these ratios. There was no reason to question the divine numerology—for followers of Pythagoras, all was number and all was perfect.

Yet even the most beautiful theory can be slain by an ugly reality. The refutation of Pythagoras's philosophy arose not from some external antagonist but from a dedicated acolyte. Little is known about Hippasus of Metapontum, but what scant records remain tell us he was a devout Pythagorean who never consciously sought to question the apparent truism of rationality.

Although there are conflicting accounts of how he inflicted such a grievous wound on Pythagorean philosophy, his work on the square root of 2 is most often cited. This held central importance to Pythagoras; consider a unit square, with each side of length 1; by his famous theorem, the central diagonal length is $\sqrt{2}$. While Pythagoreans knew this to be somewhere around 1.414, the precise ratio was not obviously deduced. They certainly tried; 99/70 gets within about 1/10,000 of the true answer. The fraction 665,857/470,832 is even better, within a trillionth of the actual answer. But mere approximation would not suffice; there had to be an exact, unique ratio for the credo to hold. Yet the search for this proved infuriatingly elusive. With an argument beautiful and ruthless in its brevity, Hippasus showed such a search to be a fool's errand. First, he assumed an irreducible ratio exists, so that: $\sqrt{2} = \frac{P}{Q}$.

Next, he banished the evasive root, and as any operation done at one side of an equality must be done on the other side, he squared both sides. After some rearrangement, this yields an equivalent expression, $2Q^2 = P^2$. At first glance, this doesn't appear to help much. But Hippasus ventured a crucial observation so seemingly trivial it might escape our notice: P^2 was twice Q^2, and thus an even number. But the only way that P^2 can be an even number is if P itself is an even number, which we'll call 2K. But then returning to our seeming trivial rearrangement, we get $2Q^2 = (2K)^2 = 4K^2$, and thus we can state that $Q^2 = 2K^2$. Using the exact same argument again, Q must also be an even number. But this can't be, as we've already defined $\frac{P}{Q}$ as an irreducible ratio—and yet the ratio of two even numbers is always reducible. Thus, an inescapable contradiction had arisen. This was an astounding conclusion—by simply assuming a perfect ratio existed, Hippasus had shown that insurmountable absurdity ensued.

The only escape from contradiction was to conclude there was no rational expression for $\sqrt{2}$, no beautiful and magical ratio. The demon of irrationality had emerged to shatter the faith, a body blow to the sanctity of the divine ratio. Worse again, meticulous application of proof by contradiction also revealed that $\sqrt{2}$ was no devilish outlier, no unique freak that could be rationalized away. Rather, it unveiled the existence of an entire new class of number, impossible to express as a neat ratio—the irrationals. And, as if to taunt the devoted, the same logic ultimately led to another revelation: the set of irrational numbers is infinitely larger than the set of all rational numbers.*

This impressive intellectual achievement did not endear Hippasus to his commune. Legends diverge on his fate for this perceived insult, and untangling the historical from the apocryphal is difficult. What is certain is that Hippasus's audacity in defiling their paradise with their own tools enraged the sect, and they convicted him of impiety. The most enduring accounts state he was sentenced to the punishment reserved for such an offence: drowning at sea. While the Pythagoreans may have killed

---

* This isn't a figure of speech; there are indeed different types of infinity. The set of natural numbers (1, 2, 3 . . .) comprise the smallest type, "countable" infinities. The set of real numbers (including irrationals) is infinitely bigger than that and "uncountable." This is way beyond our scope here, but it's an interesting thought to wrestle with. Infinities are completely nonintuitive; mathematicians refer to the smallest type of infinity as Aleph null; among its bizarre properties is the fact that Aleph null plus or minus any finite number is *still* Aleph null. This sets up a terrible set-theory joke: "Aleph null bottles of beer on the wall, Aleph null bottles of beer / Take one down, and pass it around / Aleph null bottles of beer on the wall!" The set of all number theorists has precious little intersection with the set of all comedians.

the man, they were unable to suppress the reality of what he had found. In time, the irrational brought down the very foundation of that which they considered most holy. Of course, the mathematical meaning of irrational is different from our usual definition of not being logical or reasonable. The amusing absurdity here is that the Pythagoreans' insistence on clinging to rationality was irrational when embracing the irrational was the only rational conclusion!

Contradictions are invaluable because they warn when something is askew. We're surprisingly adept, however, at ignoring them, to our detriment. Consider the fact that we are surrounded by a symphony of invisible light. Our eyes perceive but a tiny sliver of the electromagnetic spectrum, but it encapsulates every color we'll ever know and every sight we'll ever see. Electromagnetic radiation (EMR) permeates everything, from the familiar visible light that illuminates our world, to the broadcast media transmitted worldwide by radio wave, to the X-rays that have revolutionized anatomical imaging and cancer treatment. In this era of wireless communication, our phones and routers take advantage of microwave radiation to rapidly convey virtually the entire repository of human knowledge to our fingertips at staggering velocity. But in a world where mobile phones and Wi-Fi are increasingly ubiquitous, is there cause for alarm over our physical health, and might our telecommunications frequencies increase our cancer risk?

A quick glance at the internet might suggest so. Many sites vividly attest that mobile phones dramatically increase the risk of brain cancer. Others insist that our phones and routers are "cooking" us. Certain consultancy agencies stress the dangers of Wi-Fi, offering packages to minimize exposure for not-insubstantial fees. There are those too who assert that the dangers of radio-frequency radiation are being covered up by telecom giants and phone manufacturers. One such individual successfully sued the California Department of Public Health in 2017, compelling it to issue guidelines on mobile-phone radiation exposure. But perhaps the most pervasive source for these claims is the BioInitiative Report. Originally published online in 2007 to great media fanfare and updated in 2012, its bald conclusion leaves no room for ambiguity: Radio-frequency radiation is causing myriad health impacts, including huge increases in cancer risk.

Before you hurl your phone away or tear your router cable from the wall, however, it's worth noting that this stands in stark contrast to the wealth of existing scientific data. The World Health Organization (WHO) states that "no adverse health effects have been established as being caused by mobile phone use." In 2020, the FDA went on record claiming there is "no consistent or credible scientific evidence of health problems caused by the exposure to radio frequency energy emitted by cell phones." Were mobile phones causing cancer, we'd expect to see a surge in cases echoing the huge uptick in phone usage observed over the last two decades. But in huge epidemiological studies, this simply isn't seen—the 13-country INTERPHONE study found no causal relationship between phone use and common brain tumors such as glioblastoma and meningioma, with the dose–response curve betraying no signs of correlation. Nor did a large Danish cohort study reveal any obvious cancer link.

And while American mobile-phone use increased from almost nothing in 1992 to practically 100 percent by 2008, these are not non-relationships outliers, with all major studies finding zero evidence for carcinogenicity.

So why then does the confusion arise? Part of the issue is the unfortunate ambiguity of "radiation." This deeply misunderstood concept conjures up grim associations with radioactivity. This conflation is unfortunate, as radiation simply refers to transmission of energy through a medium or space. In the context of EMR, this refers to packets of electromagnetic energy moving at light speed. The electromagnetic spectrum is the range of all possible frequencies of EMR, and energy is directly proportional to frequency. While we only see a tiny portion of the spectrum in the form of visible light, we can think of it as a range of light particles (photons) with different energies.

Some of these have sufficient energy to eject electrons from atoms, smashing apart chemical bonds. This renders them capable of causing DNA damage, often a prerequisite for cancer. Light with enough energy to liberate electrons, ionizing radiation, is detrimental to our health. But even this seemingly negative property of high-energy EMR can have positive outcomes, exploited to our benefit when X-rays are harnessed to kill tumor cells in radiotherapy. This fact in isolation tends to make people uneasy, prompting a reasonable question: If light can destroy cells, could radio-frequency communications induce DNA damage and ultimately cancer?

This understandable concern pivots on misunderstanding how unbelievably vast the electromagnetic spectrum truly is. Modern communications tools like Wi-Fi and phone networks are firmly rooted at the microwave end of the scale, with frequencies between 300 MHz and 300 GHz, making them low-energy photons in the radio-frequency band. To put this in perspective, consider that the lowest-energy visible light photon (wavelength ~700nm where a nanometer (nm) is 1 billionth of a meter) carries roughly 1,430 times the energy of the most energetic microwave photon (wavelength 0.1cm). The microwave radiation that phones and routers use is indisputably non-ionizing, completely incapable of direct DNA damage. It's therefore totally unsurprising that we see no increase in cancer rates with microwave radiation, as they simply aren't powerful enough to wreak the requisite havoc on our cells.

If you're quite reasonably wondering how to square this with the apocalyptic findings of the BioInitiative Report, the short answer is you can't, because it was garbage. While masquerading as a scientific document, it was anything but. It had never undergone peer review to be rigorously assessed by experts. Media coverage and public concern brought it to the attention of scientific bodies across the world, who promptly eviscerated it. The Health Council of the Netherlands stated:

"The BioInitiative Report is not an objective and balanced reflection of the current state of scientific knowledge." Similar panning came from the European Commission's EMF-NET, the Australian Centre for Radiofrequency Bioeffects Research, the Institute of Electrical and Electronics Engineers, and the French Agency for Environmental and Occupational Health and Safety. Common to all the scientific criticism was a singular observation, voiced clearly by the German Federal Office for Radiation Protection: "[The BioInitiative Report] has undertaken to combine the health effects of low- and high-frequency fields that are not technically possible."

In the most basic terms, the report authors made an exceptionally fundamental error. To buttress their alarmist claims, they took known detrimental impacts of high-frequency ionizing radiation and presented these as if they applied to non-ionizing radio-frequency EMR, arguing that:

Premise 1:   All radio-frequency radiation is
              electromagnetic radiation.
Premise 2:   Some electromagnetic radiation can
              cause cancer.
Conclusion: Thus, radio-frequency radiation causes cancer.

This is a stellar example of the *fallacy of the undistributed middle* (*non distributio medii*), which occurs when the "middle" term in our syllogism (the term that appears in both premises but not the conclusion) is not given an explicit distribution, like "all" or "none." Here, we know that "some" EMR can cause cancer, but that isn't an explicit distribution. Conclusions drawn from this logic are inherently invalid. We can see it a little more transparently with a deliberately extreme example:

Premise 1:   All the ancient Greek philosophers are dead.
Premise 2:   Jimi Hendrix is dead.
Conclusion: Thus, Jimi Hendrix was an ancient Greek
              philosopher.

The middle term here is the state of being dead, common to both premises. Being dead isn't explicitly distributed, containing more members than just Jimi Hendrix and Greek philosophers. Without this distribution defined, the conclusion is fallacious. You could of course give it a distribution, changing the first premise to "all dead people are ancient Greek philosophers," which would render the syllogism valid but still unsound because it's a nonsensical premise. Depending on structure, the fallacy of the undistributed middle is a similar animal to either the inverse or converse errors we previously encountered. Variations on this theme are employed liberally in the roughshod world of politics. For example: "Communists favor increased taxation; my opponent favors increased taxation; thus, my opponent is a communist."

The BioInitiative Report was similarly duplicitous, conflating very different types of radiation to push a false narrative. While devoid of scientific merit, it still confounds and misleads both the public and even scientists who should know better.* In 2017, a paper published in a respected journal stated that radio-frequency EMR was not only linked to cancer, but also to autism. This paper fell across my desk, and that of psychologist Dorothy Bishop. Aside from being a wonderful writer, Dorothy is a Royal Society Fellow whose academic expertise focuses on developmental language impairments. The paper's claims about autism left her aghast, echoing my sentiments over the biophysics assertions within. The source for these woeful fictions? The BioInitiative Report, naturally. In fact, the lead author on the offending article was none other than Cindy Sage, architect of the report. Sage was not academically affiliated but managed an environmental consulting firm dedicated to reducing radio-frequency exposure—a fact curiously absent from the conflict-of-interest statement.

---

* We'll delve a little more into fears over electromagnetic radiation in chapter 11, where we'll see how aside from cancer fears, fearmongering feeds belief about a nebulous condition known as electromagnetic hypersensitivity.

The paper had garnered media interest, and Dorothy and I were approached by journalists for comment. After we pointed out flaws inherent in the work, most outlets opted not to run the story. This is not unusual—I've found that sometimes the greatest contribution a scientist can make to public understanding is to help journalists kill bad stories before they metamorphose into needless panics. Some outlets were not so conscientious; the *Daily Express* headline, with characteristic subtlety, screamed: COULD WIRELESS TECHNOLOGY BE CAUSING MAJOR HEALTH PROBLEMS IN YOUR CHILDREN?

The reviewers should of course have spotted that the only evidence proffered for such dire statements stemmed from a discredited report and rejected it outright. Instead, ineptitude somewhere along the chain had given odious fearmongering the veneer of scientific respectability and a new lease on life. In response, Dorothy and I informed the journal of this lapse in judgment with potential to cause harm. To their credit, they conceded the editorial failure, asking us to write a comprehensive rebuttal. We went slightly further, not only debunking the claims made, but suggesting guidelines for spotting the red flags of potential bad science. Sadly, radio-frequency has become even more contentious with the advent of 5G technology. Anti-5G protests have erupted worldwide, underscored by the same zombie myths which refuse to die despite repeated slaying. In 2020, cell phone towers have been hit with arson attacks and telecom engineers threatened, compounded by bizarre assertions that 5G causes coronavirus. That this is both biologically and physically impossible has proved no impediment to its adoption worldwide. This is especially damaging at a time when fears are high. And when even some scientists have been fooled by outlandish claims, or perpetuate them unwisely, that the topic still vexes people is understandable.*

---

* The conspiracy theories about radio-frequency are nothing new, yet it is interesting to see how prominent these have become in their current 5G iteration. Social media is partly responsible for this, which we'll explore in future chapters. Russian disinformation has also played a role, which we will also delve into later.

There are a handful of intimately related logical blunders that can lead us astray, such as the fallacy of *affirming the disjunct*, which assumes that two conditionals cannot concurrently be true. Let's take a trivial example: "His pet is either a dog or a mammal → His pet is a dog → Thus, it isn't a mammal." This sequence is clearly wrong, because the two options are not exclusive; dogs are a subset of mammals. If, of course, the two clauses are entirely opposed, no fallacy has been committed. For example, it is, as far as we know, impossible to be both alive and dead at once.* Accordingly, "Jimi Hendrix is either alive or dead → Jimi Hendrix isn't alive → Thus, Jimi Hendrix is dead" is not affirming the disjunct, as the two states of being cannot occur simultaneously.

The disjunct fallacy is frequently encountered in polemical form: "Either you're wrong or I'm wrong → You're wrong → Thus, I am right." This is, of course, so much bluster because the reality is that both propositions could be completely bogus. This formal error is wholeheartedly embraced in politics, where berating one's opponent is often wrongly seen as lending credence to the speaker's own position. In reality, the onus is always on the speaker to prove their own veracity, and simply exposing elements of inconsistency of other arguments—real or imagined—does not automatically validate one's own position.

Intimately related to this is drawing a positive conclusion from two negative premises—the *affirmative conclusion from a negative premise*. This is the kind of fallacious reasoning an egotistical music critic might employ for the purposes of self-aggrandizement: "I don't listen to that → People with good taste don't listen to that → Thus, I have good taste." Even if these highly subjective premises were objectively true, the hypothetical critic here hasn't justified their conclusion.

---

* This calls to mind the paradox of Schrödinger's cat, one of the most misunderstood thought experiments in modern physics. To illustrate how bizarre the quantum world was, Erwin Schrödinger drew an analogy of a cat in a box with a radioactive source that has a 50/50 chance of killing the cat. If the cat were a true quantum mechanical entity, it would exist in a superposition of being both alive and dead until observed, after which the observation would collapse it into one state or the other. Schrödinger's attempt at illustrating how strange the quantum world would be at macroscopic scales has, alas, been somewhat misunderstood. To clarify, at no stage was Schrödinger proposing that metaphysical zombie felines were a consequence of quantum mechanics.

This shoddy logic provides a foothold for those cursed with sanctimonious minds to lambaste others. Variations of this blunder are often moralistic in nature, bolstering self-righteousness at the cost of good reasoning. There seems to be a deep-seated assumption for some that attacking others for their perceived moral failing somehow implicitly sanctifies the accuser's moral position. This is nothing new—executions were once a public spectacle, with the condemned berated by the performatively pious. Such distasteful posturing has thankfully subsided in most of the world, and it might be tempting to conclude we've outgrown such pettiness. Alas, the perpetual fog of internet outrage quickly disabuses us of such a notion. Here, tedious self-righteousness inevitably follows whatever the latest perceived infraction of the moral order happens to be in vogue at the time.

Examples are depressingly abundant—such as the ridiculous persecution of Lindsey Stone. In 2012, Stone worked at a nonprofit, assisting adults with learning difficulties. Effective at her job and well liked, Lindsey shared in-jokes with friends. One benign running gag was to pose in front of a warning sign miming the opposite of the sign's direction; for instance, pretending to smoke under a no-smoking sign. Under normal circumstances, this wouldn't raise an eyebrow, a light-hearted jest with zero malice. Yet, on a trip to Arlington National Cemetery in Virginia, this innocent quirk backfired spectacularly when she posed for the camera, pretending to shout and swear beneath a sign requesting silence and respect.

Quickly, an unintended consequence of the digital age manifested—the photo, meant for a handful of friends, rapidly spread far further than intended. Stripped of the background crucial to parsing the offending picture, each share reverberated with incandescent rage. Stone had inadvertently touched a national nerve among American nationalists whose passions were inflamed by the perceived criticism of the military. As rapidly as the photo spread, so too did the outrage that anyone could disrespect the soldiers interred there. With the photo divorced from context and propagated relentlessly, Stone instantaneously became a hate figure and pariah. At least 30,000 people joined internet mobs intent on finding her, and sure enough she was located.

The chorus of shaming and abuse was absolutely frenzied. Compassion for Stone was nonexistent—not only did she lose her job, but she was inundated with threats of death or rape over her apparent lack of moral fiber. Rather understandably, she lapsed into depression and anxiety, afraid to leave her house. In the hive mind of the virtual mob was a lurking logical disconnect—something sufficiently twisted that they could justify threats of graphic violence against a harmless young woman, reveling in her downfall while still being firmly convinced they occupied the moral high ground: "She lacks moral virtue → I attack her → Thus, I am morally righteous."

Writing about the phenomenon of mobbing, scientist Joan Friedenberg remarked that "most mobbers see their actions as perfectly justified by the perceived depravity of their target, at least until they are asked to account for it with some degree of thoughtfulness, such as in a court deposition, by a journalist or in a judicial hearing." The righteousness of the mob can only be justified if the target is painted as worthy of hatred and completely crushed—and so the pursuit of warped justice is frequently animalistic, completely dehumanizing the target. As Friedenberg noted: "An unsuccessful account leaves the mobber entirely morally culpable."

Drawing an affirmative conclusion from a negative premise,* Stone's tormentors believed that the more strongly they lambasted her perceived wrongdoings, the more virtuous they were themselves. Far from being a monster, by all accounts Stone was a decent person, dedicated to helping the disabled. While there is plenty of

---

* As you might guess, there is a converse to this fallacy: negative conclusion from affirmative premises, where a negative conclusion is drawn from two positive premises. This is equally fallacious, and similarly deployed: "Either you're right or I'm right → I'm right → Thus, you're wrong."

evidence that Stone's transgression was accidental, the behavior of the shaming brigade would have been equally deplorable even if Stone had deliberately intended to offend. Even if she had been a dreadful human being, those abusing her were in no way showing themselves to be any better. The act of wielding a pitchfork does not make one heroic.

Denigrating the poor woman might have given these baying masses a sensation of moral superiority, but their conclusion was just a sanctimonious illusion pivoting on skewed logic. As with all internet storms, the raging chorus quickly forgot the human being at the epicenter and moved on to new targets of equally questionable merit. But for the unfortunate individuals who endure such totemic hate, the damage can be somewhat longer-lasting, especially when perceived transgressions needn't even be substantiated or fair to do extreme damage to the recipient.*

I'd be remiss if I didn't point out the obvious here: While logical errors underpin everything we've explored so far, there's a far more human failing implicit in much of it. We don't usually think like mathematicians or logicians, and the motivation to cling to shaky thinking often stems from something more visceral than simple misunderstanding. As we shall see in future chapters, the more strongly we hold our views, the more likely we are to accept even deeply flawed reasoning if it adds superficial clout to our worldview. We emote first, and then grasp for some intellectual justification for our initial feeling. Rather than embrace contradiction as a means to improve our ideas, we act like enraged Pythagoreans, eager to quell anything upending our comforting ideals. The sad reality is that we tend to be reactionary creatures rather than reflective ones. This is to our collective detriment because, in order to make sound decisions, we must be willing to jettison faulty reasoning—even if it sometimes means slaying our own beautiful theories.

---

* To get a sense of the damage that this causes to those at the epicenter, *So You've Been Publicly Shamed* by Jon Ronson is eye-opening.

3

IT DOES NOT FOLLOW

*Snake Oil Schemes, Cherry-Picking Psychics,
and Why Anecdotes Aren't Evidence*

We are deeply social animals and precious little influences us as much as the accounts of others. We rely upon other people's experiences, using stories and anecdotes as a psychological shortcut to index the world and all its uncertainties. Vivid accounts and emotionally charged anecdotes mold our decision-making on both a conscious level and an unconscious one. This is a double-edged sword: such illustrative accounts can help to inform our judgment and yet, by the same stroke, they can conceal or distort crucial information, rendering the conclusions we draw from them utterly false. This facet is reflected in another name for the argument from anecdote—the *fallacy of misleading vividness*, or the *anecdotal fallacy*.

Anecdotal information is incredibly vulnerable to false positives, misleading "hits" leading to a skewed impression of reality. Huge lottery wins, miraculous recovery from terminal diseases, and dramatic triumphs by underdogs make for engaging stories, but they are memorable precisely because they are unusual rather than illustrative of some underlying trend. When we infer too much from these tales, we err in our reasoning—sometimes with catastrophic results.

Let's take advertising as a clear illustration of how our innate susceptibility to personal accounts is readily exploited. Often this occurs in the form of testimonials, where customers extol the virtues of a product or service. These are extremely effective at swaying the opinions of other consumers, with word-of-mouth reviews instilling a sense of trust and coaxing new customers toward a product far more readily than a mere objective appraisal. One particularly striking example is the bizarre phenomenon of Direct to Consumer

Pharmaceutical Advertising (DTCPA), the mass marketing of medical drugs to general audiences. For ethical reasons, this practice is explicitly banned throughout most of the world. Two notable exceptions are the United States and New Zealand, where advertisements for everything from antidepressants to erectile dysfunction medications appear on television and in print, sandwiched between fashion brands and breakfast cereals.

These advertisements frequently involve patients detailing how their lives have improved since taking a given drug, or doctors extolling the virtues of the medication in question. A typical example was Pfizer's 2006 campaign for Lipitor, a cholesterol-lowering drug. In this campaign, viewers are introduced to Robert Jarvik, credited in the ads as the inventor of the artificial heart. Jarvik turns to the camera, informing us that "just because I'm a doctor doesn't mean I don't worry about my cholesterol." He tells us how the drug helped him to bring his cholesterol under control. The ad then cuts to a shot of a fit-looking Jarvik rowing across a lake. These advertisements were slickly produced, with Pfizer spending a staggering $258 million promoting Lipitor, mostly on the Jarvik campaign. They relied, however, on the viewer being unaware that Jarvik had never practiced medicine in his life and would not have been professionally allowed to prescribe any drug.

When the House of Representatives' Energy and Commerce Committee began investigating, Jarvik was forced to admit that he had not taken the drug prior to becoming company spokesman. To compound matters, his former colleagues at the University of Utah went on record to assert that he was not even the inventor of the artificial heart, an honor they claimed belonged to Willem Kolff and Tetsuzo Akutsu. Amid the controversy, Pfizer eventually dropped Jarvik in 2008, but even so the campaign was highly effective; a study by the Consumer Reports National Research Center showed the campaign sustained Lipitor's position as the number one cholesterol-lowering drug, with 2007 sales of $12.7 billion. Further, 41 percent of viewers were convinced that Lipitor was better than generic alternatives, even though these equally effective drugs were available at half the cost. Most tellingly perhaps, 92 percent of respondents liked the ad, finding Jarvik convincing. In a bizarre footnote, it was revealed that Jarvik hadn't even been rowing the boat across the lake in the ad's most picturesque scene; the advertising agency had opted for a more athletic body double.

Such events are not rare. In the Information Age, where online reviews are standard on most trading sites, false reviews and planted testimony are a persistent problem. Online commerce is plagued by dubious testimonials. A fertile cottage industry exists to write bogus flattering reviews. Such is the power of testimonials that trading standards organizations in many countries have been forced to intervene; the American Federal Trade Commission introduced legislation in 2009 to combat fallacious testimonials. However, given the sprawling nature and tangled jurisdictions of the internet, such violations are difficult to police.

Online ratings systems are also notoriously easy to game. In 2017, The Shed at Dulwich became London's highest-rated restaurant on TripAdvisor. Its stellar reviews saw it occupying first place among the capital's 18,149 rated dining establishments, with London's finest scrambling to get a seat. Unbeknownst to them, the Shed didn't exist—it was a hoax by writer Oobah Butler, inspired by his experiences of writing paid reviews of restaurants in which he had never even set foot. The principle of *caveat emptor* should be maintained, even in the face of fawning praise.

This is hardly a new problem; for as long as humans have become ill, people have sworn on the curative powers of magical elixirs, from asses' milk to bear bile. Throughout history, where there has been suffering there have been charlatans only too happy to exploit it for profit. The sheer ubiquity of odious cranks throughout the ages is exemplified by the abundance of terminology for such individuals, all with diverse etymology. The French word "charlatan" dates from the 1600s, referring pejoratively to one who peddles a medicine with elaborate theatrics. "Quackery" is at least two

centuries old, derived from the Dutch "Quacksalver"—one who
hawks salves.

The term "snake oil" is today understood as a derogatory reference to the wares peddled by purveyors of fraudulent or unproven
medicine. But snake oil originally referred to a concoction derived
from actual snakes, which came to prominence during the construction of the first transcontinental railroad, linking Iowa to
California, between 1863 and 1869. This mammoth undertaking
involved bringing workers from all from over the world to America
to lay more than 1,900 miles (3,000 km) of iron track. This was
back-breaking work and, unsurprisingly, many workers were
afflicted with sore joints. Folk remedies were eagerly traded. The
international workforce also boasted a substantial Chinese contingent among its ranks, who swore by a traditional, easily obtained
cure-all: snake oil. When they traded the oil with their American
colleagues, tales of great improvement abounded.

These tales spread rapidly, with the alleged benefits multiplying
in the retelling. Keen Western entrepreneurs saw a market, and
rapidly an empire of hucksters sprang forth. Armed with theatri-
cal flamboyance and Barnumesque showmanship, they relied on
breathless testimonials from audience plants to increase the excitement,* doing a roaring trade despite none of the evidence rising
above the anecdotal.† One of the finest snake oil salesmen was the
self-declared "Rattlesnake King," Clark Stanley, a fraudster with a
suitably ludicrous back story. Clark claimed to have spent 11 years as
a cowboy, during which time, in exchange for a demonstration of
his shooting skills, he was taken in as an apprentice by a mysterious Hopi medicine man in deepest Arizona. It was during these
studies, Stanley claimed, that he learned the miraculous powers
of snake oil and, assisted by a Boston druggist, he began selling his

---

* Audience plants waxing lyrical about the elixir in question were the "shill"—in such cases, the
term is apt.

† If such remedies are ineffective, why did they have such a draw? Part of the answer lies in regression to the mean, and our innate human susceptibility to expectations, all of which we'll cover later
on. In the case of the railroad workers, lack of alternatives might also have played a significant role.

wares in person to enraptured audiences countrywide. These audiences were not insubstantial; the Chicago World's Fair of 1893 was the biggest public event ever held in America—and there was Stanley, milking the role of swashbuckling frontiersman, dazzling onlookers by killing rattlesnakes before their very eyes. He'd then squeeze the snakes' lifeless bodies, proclaiming that the fluid that emerged was a magical elixir, panacea to a trove of ills.

Business boomed. At one stage, Stanley had several premises and boasted of killing up to 5,000 snakes a year in order to keep up with demand. But the golden age of the shyster was fading, and in 1906 the US government brought in the Pure Food and Drug Act to stem the tide of fraudulent cure-alls. Nonetheless, Stanley continued to work, until in 1916 analytical chemists subjected his much-lauded panacea to rigorous analysis and found that it chiefly consisted of somewhat more earthly ingredients: mineral oil and turpentine. There was not a drop of snake oil in Stanley's marvellous medicine. Fined $20 for misleading advertising, Stanley quietly faded into obscurity while, in time, snake oil became the catch-all term for any supposedly miraculous elixir. Even today, in a world of stricter drug enforcement and trading standards, the market for metaphorical snake oil remains solid, with miraculous cures for every imaginable ailment doing a roaring trade, buoyed by gushing testimonials and armies of true believers with anecdotes aplenty.

At this juncture, it's worth stating explicitly something that we've only implicitly touched upon until now. The underlying problem with all formal fallacies is that somewhere in the argument lurks a misstep in logic that renders the argument invalid. This means that all the formal slips are non sequiturs (literally "it does not follow"), where a conclusion doesn't flow from a premise. Any non-sequitur leap in logic constitutes an inherently false argument. Conclusions drawn from anecdotes are especially problematic—after all, with snake oil cures, it might be tempting to infer from the positive accounts that there must be some merit to the treatment. At best, many cure-alls will be nothing worse than ineffectual. But at the darker end of the spectrum they can be actively harmful, either directly or by staving off necessary medical intervention.

Vivid stories also drive panics and fuel epidemics, not least because the involvement of real people captures the imagination far more than mere statistics. There is a very understandable human tendency to focus on graphic specific cases while neglecting to account for the underlying base-rate information that might give us a better grasp of how illustrative–or extraordinary–the specific case is. This is called the *base-rate fallacy*, a term that refers to our propensity to jump to conclusions from a single example without an appreciation of the underlying reality. Anecdotes can exacerbate this problem, disproportionately capturing our attention.

This isn't always obvious–observation stripped of context can subtly encourage non-sequitur leaps in reasoning. Consider, for example, the fact that cancer incidence rates have increased considerably through the twentieth and twenty-first centuries. This is undeniably true; for much of the twentieth century, roughly one in three people developed a form of cancer. More recent estimates show this proportion has climbed considerably, with half of us likely to develop cancer in our lifetime. This is alarming, and in the scramble to find a scapegoat for this, many are quick to attribute blame to everything from genetically modified food to vaccination.* Even the more scientifically informed might struggle to find a culprit.

But our environment hasn't become more toxic. We're simply living longer. In fact, cancer survival rates have never been higher and continue to climb due to improved diagnosis and treatment. Cancer is primarily a disease of aging, with age the single biggest risk factor. Having largely circumvented the litany of infectious diseases, poor sanitation, and abject pestilence that plagued our forebears, we're now living longer. The apparent rise in cancer rates is paradoxically a symptom of improved societal health. Yet

---

* There are a host of misguided and devious characters who exploit such base fears rather profitably–we'll encounter some of them in future chapters.

the fact that cancer rates have increased in isolation can encourage a jump in logic completely at odds with the reality of the situation.

There's an important caveat: if anecdotes are so frequently dubious, how do we acquire data? The plural of anecdote, scientists warn, is not data. What does that mean? After all, an anecdote, if accurately reported, might give us some insight into possible outcomes of a system. For example, we know that people do win the lottery. What anecdotes alone cannot tell us is whether such outcomes are representative or common. Sometimes, the issue is simply that the information we really need is initially invisible, and the available anecdotes mask the true situation. An especially illustrative example comes from the Second World War, when American and Japanese pilots vied for supremacy of the Pacific sky. Deadly high-altitude dogfights were a constant feature, with high losses on both sides. To stem the bloody tide, the Center for Naval Analysis (CNA) decided to study the bullet-riddled fuselages of returned fighters to ascertain their weak points.

Analysts pored over the data from damaged planes, mapping the extent and location of the damage. Hits were distributed all over the hull, with some areas—like the engines and cockpit—oddly spared major scarring. Given the paucity of case studies exhibiting damage around the cockpit, the engineers opted to leave them and fortify other regions. But one statistician, a man named Abraham Wald, realized this absence was important and instead told a completely different story. The reality was that the fighters who took damage to their engines and cockpit went down in a fiery haze, and therefore never made it to analysis. This insight tore the CNA's painstaking work apart, leading to the opposite conclusion.

Such errors are known as *survivorship bias*, when one inadvertently overlooks cases with a lack of visibility and instead bases conclusions only on successes. In highly competitive careers, this often manifests as the underdog story—the billionaire CEO who dropped out of school or the self-trained musician who made it big. The implicit lesson there is that anyone can make it, but this ignores the huge role of luck and timing. It turns a blind eye to the multitudes of similarly talented people who fell by the wayside in these professions—the human equivalent of Abraham Wald's missing cockpit data.

There is another exercise in dubious logic that is intimately related to the anecdotal fallacy, and sometimes indistinguishable from it. If the anecdotal fallacy is the vehicle, then the *fallacy of incomplete evidence*, or the *cherry-picking fallacy*, is the engine that drives it. Cherry-picking is the selective use of evidence to reject or ignore details that contradict or undermine the speaker's assertion. The evidence in question might vary in type and scope; it may be the carefully curated selection of supportive anecdotes and testimonials we have previously seen in the anecdotal fallacy. Worse, it might be a selective fixation on only the data that jibes with one's prejudice while ignoring what the evidence truly conveys.

This is a severe problem in public discourse and a persistent pitfall in communicating science to the public. On topics as diverse as alternative medicine and climate change, those with vested interests can attempt to circumvent the scientific consensus by clinging to outliers from the noisy data of our world, even when the quality evidence and analysis is against them. What we need to understand is that not all evidence or experiments are created equal, and that it can take sophisticated tools and methods to unravel causal relationships. Cherry-picking is the mechanism that sustains belief in the face of overwhelming evidence against it. And to see that in action, we need only look at the popularity of psychics.

Fantastic stories of psychic ability are common and there are all manner of tales describing apparently uncanny psychic skills. However, if we subject these stories to a meticulous analysis, a different picture emerges. A classic example came in 1997, when Richard Wiseman and Donald West undertook a study pitting undergraduate students against self-proclaimed psychics by giving both groups objects from solved crimes and asking for the details of that crime. The psychic group in the study performed no better than the undergrads, and neither group performed better than chance alone. And this is not a mere outlier anecdote, for under stringent testing conditions no psychic has ever demonstrated a plausible ability. Indeed, the National Academy of Sciences stated in its 1988 report on the subject that there exists "no scientific justification from research conducted over a period of 130 years for the existence of parapsychological phenomena."

Yet the market remains buoyant. In the US, approximately 60 percent of survey respondents agreed with the statement "some people possess psychic powers or ESP," while an estimated 23 percent of the UK's population have consulted a psychic. A quick Google search yields an absolute slew of mediums, clairvoyants, and psychics jostling for business on websites and telephone hotlines. So how do they persuade us? The answer, it appears, is that the counter-intuitive popularity of psychics is based entirely upon carefully curated anecdotes and a great amount of cherry-picking. Psychics will parade their hits while downplaying their misses, fostering an illusion of clairvoyance. As we shall see, psychics are also aided by the psychological blind spots and statistical innumeracy of their devotees, but naked cherry-picking is a key element.

Those who are especially good at cherry-picking or manufacturing impressive-seeming anecdotes can establish profitable careers. This might seem a harmless indulgence, and a cynic might mutter something about a fool and his or her money, but many psychics profit from the bereaved and the anxious. People disproportionately visit psychics when in a state of uncertainty. While the advice dispensed might be little more than a truism, it can also be actively damaging. Consider Sylvia Browne, who from 1974 until her death in 2013 was perhaps the best-known psychic in America. A darling of daytime television, Browne's most significant gift was her talent for shameless self-publicity, claiming an uncanny success record and taking credit for solving several high-profile crimes as a police consultant. In reality, of the 35 police cases Browne involved herself in, the information she provided in 21 was too vague to be useful. In the remaining 14, police officers and family members insisted that she had played no useful role.

Browne seemed unimpeded, continuing to insert herself into high-profile missing people's cases on national television. When Gwendolyn Krewson's 23-year-old daughter Holly disappeared near San Diego in 1995, for instance, her distraught family turned to Browne in desperation. On national television, Browne confidently predicted that Holly was alive and working as a stripper in Hollywood. With renewed hope, the distraught Krewson family began an intense search that spanned years, yielding absolutely nothing. Holly's remains were finally identified from dental records on a Jane Doe in 2006—a body that had been discovered almost a decade before. The autopsy revealed she had died soon after her disappearance. Browne's convoluted tale had absolutely no merit.

And so it continued. When Lynda McClelland went missing in 2002, for example, Browne, on daytime television, told Lynda's son-in-law, David Repasky, that she had been kidnapped by a man with the initials "MJ" and would be found alive soon. In 2003, McClelland's body was discovered two miles away and forensic examination revealed that she had been murdered by none other than Repasky himself—a slight oversight. Utter callousness was frequently displayed to terrified and desperate relatives. When six-year-old Opal Jo Jennings was snatched in front of her grandparents' house near Fort Worth in 1999, Browne coolly stated that she was alive, but had been kidnapped and forced into prostitution in a town called Kukouro in Japan. When Opal's remains were eventually discovered in 2004, it was shown that she had died of blunt-force trauma just a few hours after her abduction. Don't expend too much energy looking for Kukouro either—no such place exists.

With each audacious move, Browne's profile—and profitability—grew. She became a regular fixture on the syndicated *Montel Williams Show*, making grandiose predictions, and eventually was earning over $3 million a year, charging $750 for a 20-minute phone consultation. She continued her heartless excursions into emotional manipulation, surrounded by doting audiences convinced of her skill. Browne's "knowledge" also extended to the world of medicine. There exists some eyebrow-raising footage of Browne telling a woman in pain after surgery that a metal implement had been left inside her and that she needed to have a full-body MRI. This might have seemed reasonable, until one considers the tiny detail that an MRI machine is essentially a huge magnet that could have ripped any metal out of a body. In Browne's defense, that would at least have removed it.

These are but a few examples from a career defined by jaw-dropping inaccuracy. And while Browne claimed an 87–90 percent success rate, analysis of her performance on the *Montel Williams Show* gave her a success rate of precisely zero. Browne shrugged such criticism off with the line "Only God is right all the time"—a nonchalance indicating that Browne's neck was so brass, one could polish it and call it a door knocker. Her high profile, coupled with the sheer audacity of her baseless pronouncements, invoked awe from her admirers and ire from skeptics. At one stage she asserted she would take up the skeptic investigator James Randi's offer of a million dollars for successfully demonstrating her ability. She further asserted she would readily win, yet persistently found excuses for avoiding the test right up to her death. In keeping with her character, Browne confidently predicted that she would die at the age of 88, but instead died in 2013 at the age of 77—one final bungled prophecy.

Sadly, Browne is far from alone. The TV psychic John Edward faltered when examined by James Randi. Derek Acorah, of *Most Haunted* fame, was famously duped by psychologist Ciarán O'Keeffe into communicating with the spirit of "Kreed Kafer," an entirely fictional person who just so happened to have a name that was an anagram of "Derek-Faker." In the UK, the notoriously litigious psychic Sally Morgan is a household name, dogged by accusations of fakery. After a less than fawning article, Morgan attempted to sue the magician Paul Zenon for £150,000 (about $200,000) in damages.

The president of the James Randi Educational Foundation (JREF), D. J. Grothe, questioned precisely why Morgan would sue for such a paltry amount when she could readily have claimed millions of dollars from the foundation were her powers real. As he wryly remarked, "it makes one wonder if even Sally Morgan believes that Sally Morgan's powers are real."

Yet the bank balances and audience sizes of these individuals remain healthy. But if psychics perform no better than chance, why is there such a market for their services? A large part of the answer lies in cherry-picking. Random predictions will occasionally be right, and psychics focus on these, emphasizing their hits while downplaying their losses. The hits can be surprisingly easy to ascertain—one age-old trick employed by mentalists and psychics alike is cold-reading, where a reader can quickly ascertain seemingly ethereal knowledge simply by analyzing the subject's body language, visual cues, clothing, age, or manner of speaking. Typically, these guesses have a high probability of being true, and when they are hits the performer will seize upon them, glossing over those that don't succeed. Done correctly, this gives an impression of prior knowledge and the psychic can then use all the positive "hit" anecdotes to cement the illusion further.

The psychic can also hedge their bets by using the rainbow ruse technique, which involves uttering a platitude that endows the subject with two opposite traits simultaneously. For example: "You're generally a positive and upbeat person, but there have been times when you were very sad." This is likely to be agreed upon by practically everyone, providing yet more apparent anecdotal evidence to a suggestible listener or audience. Most common is a related technique known as shotgunning, which relies on rapidly projecting copious amounts of very vague information, hoping to hit a target and evoke a response. A shotgunning statement might be something like: "I see a man, who passed with a heart problem; perhaps a father or father figure . . . a grandfather, an uncle, a cousin or brother . . . I am clearly seeing chest pain here." With a reasonable audience, listing a number of male figures is bound to garner hits, not least because roughly half the male population in the Western world die from heart problems.

This is the crux of the problem; psychics are supported almost entirely by the siren-song of cherry-picked "hits" that they encourage their audience to focus on. But these are illusions—no matter how seemingly impressive the hits, it is a complete non sequitur to infer paranormal ability. Similarly, testimonials may tempt us to believe in the impossible or unlikely, but we err in our reasoning when we allow such empty tactics to sway us. If we want to assess whether a medicine works or whether a particular course of action is optimum, we mustn't rely simply on favorable accounts.

Over the past few chapters, we have covered some of the most common errors we make with the structure of a logical argument. There are, of course, more arcane and convoluted forms that occur in an argument, but these are the most regular offenders. To this point, we've concentrated on arguments rendered intrinsically void due to some glitch in the underlying logical structure. In more formal terms, we have focused on the validity of a line of reasoning. But for an argument to be sound, it requires more than just a valid syntax. In mathematical logic, an argument can only be sound if both the structure is valid and the premises are true.

Yet it is not only blunders in logic that lead to dubious conclusions—as we saw with our time-traveling-robot Greek philosopher example, the structure of an argument might be logically reasonable, but if the premises are flawed then the conclusion is questionable. Such errors are known as *informal fallacies*. And as there can be much ambiguity in premises, these can be used as a rhetorical Trojan horse to shuttle in all manner of dubious conclusions. The underhandedness of informal fallacies is such a rich and important area that it is vital to detect their malign influence. Accordingly, we'll dedicate the next few chapters to his topic.

But before we move on, there is a vitally important lemma (a logical stepping stone) here that shouldn't be overlooked: The mere fact that an argument contains a logical fallacy does not necessarily render the conclusion incorrect. Ironic, isn't it, that proclaiming a conclusion is incorrect, solely because an argument behind it is wrong, is itself a non sequitur? It is entirely possible to be right for the wrong reasons, and a poorly argued proposition does not always render a claim wrong. This error is *argumentum ad logicam* (argument to logic) or the *fallacy fallacy*. To take an outlandish example, imagine if your friend believes that you shouldn't put your hand in the fire because they once did that and subsequently lost their keys. While you would be correct in dismissing their argument as a bizarre non sequitur, presuming their conclusion is bogus might be unwise, as sticking one's hand in the fire is generally not a good course of action unless one just so happens to be fire retardant.

This is something of which we need to be mindful: It can be relatively easy to dismantle an argument, but it can take rather more finesse to evaluate the claim independently of the sound and fury surrounding it. In other words, valid conclusions can be wrapped in bad reasoning. This only becomes more pronounced when we consider how rhetoric misleads us. To see this, we'll next dip our toes in the deep ocean of informal fallacies and explore the devious ways in which these glitches in our reasoning can be manipulated.

PART 2

The Pure and Simple Truth?

Spotting and Debunking Dubious Rhetoric

"The pure and simple truth is rarely pure and never simple."

—OSCAR WILDE

4

THE DEVIL IN THE DETAILS

*Reductive Cause Fallacies, Why Vitamins Aren't Panaceas,
and the Perils of False Dichotomies*

I f you've ever labored under the misery of a cold, then you've likely
heard a well-meaning friend recommend vitamin C to stave off
ill effects. This enduring belief owes its popularity to an unlikely
figure–the towering intellect Linus Pauling. A polymath of note,
Pauling's interests spanned everything from quantum chemistry to
DNA structure. His accomplishments are no less impressive; to this
day he remains the only person to have received two unshared Nobel
Prizes, one for Chemistry in 1954 and the Peace Prize in 1962. Fellow
Nobel laureate Francis Crick, the co-discoverer of DNA, lauded
Pauling as the "father of molecular biology." During a lecture in the
1960s, Pauling spoke of his desire to live at least another 25 years so
that he might keep abreast of advancements in science. This might
have remained a throwaway remark had a man named Irwin Stone
not been sitting in the audience. Soon after, Stone penned a missive
to Pauling recommending what he claimed was the elixir of vitality:
3,000 mg of vitamin C every day.

A more cynical man might have dismissed this advice as deeply suspect or outright quackery. Pauling displayed no such caution, opting
to follow Stone's regimen. Immediately, he reported feeling imbued
with more energy and even less susceptible to colds than before.
Hugely enthused, Pauling increased his dose steadily over the years
to a staggering 18,000 mg a day. By 1970, he had become something
of a zealot, authoring the first of his tomes on the subject, *Vitamin C
and the Common Cold*, extolling the virtues of mega-dose vitamins.
The book was a huge success, and overnight people began buying
huge quantities of vitamin C, convinced it could stave off colds. In
some places, sales increased tenfold within a year to the extent that
pharmacists couldn't keep up with demand. The reassuring message

that vitamin C could circumvent the drudgery of illness was taken to heart across America and the world—after all, this was medical advice from a multiple Nobel Prize winner.

But Pauling's evangelical zeal was not well grounded in evidence. Aside from a handful of anecdotes, there was simply no convincing rationale that mega-doses of vitamin C had any tangible benefits. A scathing review of Pauling's book for the *Journal of the American Medical Association* in 1971 by physician Franklin Bing took Pauling to task for making claims not buttressed by evidence, lamenting that "unfortunately, many laymen are going to believe the ideas that the author is selling." Bing had no idea how right he was—nor how enduring the myth would prove. Subsequent studies found precious little evidence to support Pauling's central thesis, with doses of up to 10,000 mg performing no better than a placebo. Undeterred, Pauling increased his claims in magnitude and scope. He published more books on the subject, insisting vitamins were a universal panacea for everything from cancer to snakebites to AIDS.

Even as more evidence rolled in that his contentions were mistaken, Pauling remained steadfast in his conviction, confidently predicting that those who used a high-dose vitamin regimen would live up to 35 years longer, free of diseases. Pauling eventually died in 1994,* but his ideas on vitamin C endure to this day and show precious little sign of abating. Indeed, far from being beneficial, mega-doses of vitamin C are not encouraged. Side effects of large doses can include severe flatulence and diarrhea, prompting the more scatologically minded to wonder if the increased activity Pauling reported was perhaps chiefly in his bowels. But what is undeniable is that Pauling's perceived authority is a large part of why the myth took root in the first instance. Far from fading into obscurity, the myth thrives to

---

* Pauling himself died at 93, but vitamin C had little to do with his advanced age; good health care and lucky genetics are a more likely bet. On a tangent, I had the pleasure of pathologist Sir Michael Epstein's company at a Wolfson College Fellows dinner in Oxford in 2016. Epstein was 95 years of age at the time, unbelievably sharp and in great physical shape. When one of the other fellows asked to what he attributed his health and longevity, Epstein replied with a grin: "The secret is to choose the right parents."

this day, with the even more extreme Intravenous Vitamin Therapy growing in popularity, where vitamins are delivered via drip. As we shall see in chapter 18, this trend is largely due to celebrity endorsement, but aside from having zero medical efficacy, this can be extremely dangerous. In 2019, 29-year-old Russian model Janna Rasskazova died while on IV vitamins, while a 63-year-old grandmother in the UK died from the procedure in 2020.

The endurance of such myths stems in part from how we understand terms and concepts. Of all our defining characteristics, language is perhaps the most unique and potent. At the dawn of humanity, we were endowed with just the right evolutionary quirks to give us the physical apparatus to speak and the mental capacity to transcribe thoughts into words. This ability is central to what it means to be human, yet language is fraught with ambiguity and equivocation. The words we use are rich in meaning, frequently impossible to isolate from the context in which they're uttered. This ambiguity gives us the tools to convey rich and nuanced meaning; our poetry, humor, and theater thrive on language's glorious amorphous nature. The protean quality of language, however, can cloak a variety of sins, and flexibility of our words and concepts sometimes makes it all too easy to be led astray. The nebulous nature of certain concepts is ripe for confusion, and "expert" is certainly one such word.

We often defer to the wisdom of experts to guide our judgments. For example, we generally bow to the advice of a physician on medical matters, an eminently reasonable position to take given the extensive training doctors undergo. But the situation isn't always so clear-cut and, as Pauling demonstrated, expertise in one domain does not translate to expertise, or even sanity, in another. An *argument from authority* is where a perceived authority's support is used to justify a conclusion. But there is a serious and often insurmountable problem with assuming the inerrancy of an expert. Politicians, for example, might be experts on aspects of policy and democracy, but their judgments will differ wildly with ideological positions. Even ostensible authority might itself be contentious, as "expert" is a nebulous term, difficult to pin down. If the question concerned an ethical dilemma, for example, then depending on the individual in question the expert might be a priest or a philosopher, each of whom will likely give differing advice.

Even in our medical example, there is room for subjectivity. While trusting a medical doctor is usually a justified assumption, purveyors of alternative medicine tend to speak with authority, even without evidence for their claims. Some qualified physicians subscribe to unproven or debunked ideas. Ostensible authorities can fall victim to error through inadequate knowledge, bias, dishonesty, or even group-think. Relying solely on authority is frequently treacherous, especially when the expertise at hand may be inherently questionable—the predictions of economists, for example, often conflict despite their learning.*

The argument from authority is a classic informal error. These arise when something is remiss with the premises of an argument, even if the logic is valid. Precisely what is wrong varies greatly—the premises may be too weak to support the conclusion, or the language too equivocal, or the generalizations faulty. Just as language gives us a staggering multitude of ways to express ourselves, so too does it open up equally abundant gaps for dubious inferences to lurk. Arguments from authority tend to rely on a monolithic, unchanging interpretation of expertise. But this approach breaks down when the frontiers of knowledge are themselves in rapid flux.

In the 1840s, the Hungarian physician Ignaz Philipp Semmelweis took up an obstetrics post in Vienna General Hospital. Around the time he arrived in Vienna, a surge of illegitimate pregnancies had led to several infanticides across Europe. To stem this grisly trend, free maternity hospitals hastily opened across the continent. Vienna General was home to two such clinics. This was a time long before

---

* Playing devil's advocate, might one contend that science itself is a mere argument from authority? As we'll see later, the answer is no, although we'll also see how individual scientists have on occasion exploited public trust to push falsehoods.

antibiotics, and childbirth was still an inherently risky undertaking. Many mothers died of infection soon after birth. Curiously, despite the two clinics in Vienna being similar in almost every respect, the first clinic had much higher mortality rates than the second. This fact was known to the expectant mothers, who begged to be hosted in the second, preferring to give birth on the street to avoid the dreaded first clinic.

This inexplicable difference in mortality intrigued Semmelweis but, initially at least, his investigations amounted to little. During a mundane autopsy in 1847, his colleague Jakob Kolletschka accidentally nicked himself with a scalpel and subsequently fell fatally ill. The macabre spectacle of his demise yielded a morbid clue, as his agonizing symptoms were identical to those of the women to whom Semmelweis helplessly attended each day. In the wake of Kolletschka's death, Semmelweis conjectured that decaying organic matter—a cadaveric particle that transmitted disease—led to the infections. His notion was bolstered by a difference between the clinics he had until this point overlooked: Physicians at the first clinic practiced their surgical acumen on corpses when not attending to the women in their care. To test out this theory, Semmelweis instituted a strict hygiene regime to rid doctors of these cadaver particles, including chlorine washes to remove the stench of death. Almost instantaneously, death rates plummeted. Within a month, mortality rates at the first clinic matched the safer second one, and fever deaths hit unprecedented lows.

Despite the undeniable success of his experiment, Semmelweis found himself antagonistically opposed by much of the medical establishment. Medicine in the 1800s stood on the cusp of the scientific era, but older physicians still preached the ancient concept of humorism, the belief that all disease was caused by imbalance of the four "humors"—blood, yellow bile, phlegm, and black bile. Under this schema, the primary role of a doctor was to adjust this imbalance by employing techniques like blood-letting. Many physicians were still immersed in teachings of antiquity and had at best a passing acquaintance with the fledgling scientific method. Instead, they relied on knowledge passed down by seniors and professors. This rendered medicine a field rife with strong personalities and forthright opinions, with many interventions more rooted in ritual than evidence.

Semmelweis's claims not only flew in the face of this school; they offended the sensibilities of many physicians, who were outraged by this young upstart's implication that they might be unclean. Many simply dismissed his work, asserting that if one is at odds with the authority of medicine, then it follows that one is wrong. By 1865, Semmelweis was driven to utter distraction by the lack of enthusiasm over his work. Already showing signs of cognitive impairment, he began drinking to excess and penned a series of vicious letters to his critics—each more inflammatory than the last. Embittered and angry, he denounced obstetricians as "irresponsible murderers" and "ignoramuses." This behavior took its toll on both his professional standing and the potential acceptance of his data from scientifically inclined, less dogmatic peers. At the age of just 47, he was committed to an asylum.

Mental health of the era was even less illuminated by the light of scientific inquiry. Semmelweis was bound in a straitjacket and doused with cold water. After attempting to escape, he was beaten so savagely that a wound set in. Within two weeks, Semmelweis ironically succumbed to infection, dying out of sight and mind. His funeral was poorly attended, save a few scattered family and friends. In Vienna General, his successors dismissed his findings with contempt, safe in the knowledge that their authority meant Semmelweis was mistaken. Rapidly, death rates rocketed back up needlessly. It took decades for the observation that hand-washing saves lives to be commonly accepted, with countless young women paying the price for this folly.

Semmelweis's ordeal is frequently cited as a clear case of argument from authority. This is true to an extent, but contemporary reaction to Semmelweis's discovery was more nuanced than popular accounts often state. Semmelweis was far from the first to suggest hand-washing in lime, though his studies on the subject were indeed valuable. His theory of all disease arising from cadaveric particles was, however, completely incorrect. It would still be years before Louis Pasteur revealed the existence of microorganisms, but even during Semmelweis's lifetime there was abundant evidence proving his universal model of disease was evidently incorrect. There was also confusion over precisely what he had discovered, a fact not helped when he became increasingly unhinged. Even the emerging scientific contingent was dismayed by his insistence that there was only one cause of disease, and his denial of airborne infection—both demonstrably incorrect positions, even then. Nor did his strident (if sometimes understandable) reaction to his critics endear his work to others.

Given all this, you might reasonably ask why I have included it. The answer is twofold: First, the narrative is well known and there is some evidence that Semmelweis at least sometimes ran afoul of authority. Second, and more important, the nuance around Semmelweis's story rather beautifully illustrates an even more dangerous and persistent flaw in reasoning. His chief scientific blunder was to attribute the complexity of multifaceted diseases to a single cause. His insistence on seeing everything through this misshapen, reductive lens rendered many of his conclusions wrong. Semmelweis's story inadvertently demonstrates another frequent and damaging blunder: the *fallacy of the single cause*, or the reductive fallacy.

A desire to find universal causes for things is understandable. We have an intrinsic desire for simple narratives, where cause and effect are clear and well defined. Yet, in the interwoven machinery of reality, this is often the exception rather than the rule. Perhaps because of our yearning for an overarching, easily grasped narrative of the random motion of life, single-cause fallacies are appealing but usually completely wrong or so reductive as to be useless. Nevertheless, our lust for finding meaning in noise means we constantly employ this approach even when it is vapid in the extreme. It is used to the point of tedium in political and media discourse, where ideologically driven pundits offer simple explanations and solutions for complex phenomena, seemingly without cognizance of the fact that many situations have concurrent causes and contributing factors. The reductive fallacy can serve as a vehicle for the most noxious of social and political fictions, and the associated cost can be quite simply staggering.

In 1918, at the twilight of the First World War, the highest echelon of the German army, the Oberste Heeresleitung (OHL), was a de facto military dictatorship. By the end of the spring offensive on the Western Front, it was clear to the high command that the war was all but lost. Seeing inevitable defeat on the horizon, the OHL rapidly implemented a transition to a rudimentary parliamentary system. Under this new civilian authority, a peace accord was reached and the war ended. But the armistice of November 1918 threw nationalistic right-wing elements of the German establishment into disarray; how could the might of the imperial war machine have been so thoroughly overcome? Their shame was compounded by the terms of the Treaty of Versailles, which laid blame for the conflict firmly at German feet.

The break-up of the once-proud German military and navy and the stiff financial cost the failed war effort incurred were deemed incredibly humiliating by the militaristic contingent of the German empire. Many of them simply refused to even countenance the multitudinous factors shaping German military decline. From the ashes of wounded national pride and the complex realities of a bloody war arose a terrible myth: The German defeat must be due to traitorous elements on the home front who had conspired to destroy Germany from within. The myth was adopted wholeheartedly by many, even those who should have known better, such as General Erich Ludendorff. When dining in 1919 with the British general Sir Neill Malcolm, an impassioned Ludendorff reeled off a rambling litany of reasons why the German army had been so thoroughly routed the year prior. In this frenzy of excuses, he dropped the now-infamous canard that the home front had failed the military. Historian John Wheeler-Bennett recounts the conversation between the two military men:

> Malcolm asked him: "Do you mean, General, that you were stabbed in the back?" Ludendorff's eyes lit up and he leapt upon the phrase like a dog on a bone. "Stabbed in the back?" he repeated. "Yes, that's it, exactly; we were stabbed in the back." And thus was born a legend which has never entirely perished.

Following this fallacious epiphany, Ludendorff became the leading evangelist for the *Dolchstoßlegende*, or stab-in-the-back myth. This convenient fiction placed the blame squarely on the shoulders of lurking saboteurs, and was adopted eagerly by many in German society. The identity of these nefarious elements varied with the prejudices of the believers: Bolsheviks, Communists, pacifists, trade unionists, republicans, Jews—sometimes combinations of all these detested types. It resonated with ultranationalists, echoing the symbolism in Richard Wagner's opera *Götterdämmerung* of Hagen burying his spear in Siegfried's exposed back. The early democratic leaders of the Weimar Republic and signatories of the German armistice were denounced as the "November criminals'"by rabid right-wing reactionaries. These feelings ran angry and deep; signatory Matthias Erzberger was assassinated by the ultranationalistic Organization Consul in 1921, with foreign minister Walther Rathenau murdered by the group the following year.

Of course, the simplistic betrayal explanation was devoid of any substance, thoroughly refuted by scholars both inside and outside Germany. But a complete lack of veracity is rarely an impediment to an easily grasped story's taking firm hold. Believers in the myth cherry-picked alleged instances of "betrayal" by rogue elements. For instance, Kurt Eisner, a Jewish journalist, was convicted of treason for inciting a strike at a munitions factory in 1918. Eisner himself was assassinated by a nationalist the following year. As Ludendorff must have known, such actions were inconsequential to German defeat. By 1918 Germany was already out of reserves and for an array of reasons completely overwhelmed. But admitting that Germany's defeat had several complex influences didn't jibe with the same reassuring simplicity that the stabbed-in-the-back narrative provided. The legend gave believers something else too: a scapegoat for perceived failings. From this face-saving fiction something even more poisonous emerged: a virulent new strain of anti-Semitism and deep-seated political hatred. This twisted alternative history found a charismatic mouthpiece in the form of a young Austrian firebrand named Adolf Hitler.

Hitler embraced the myth completely, fusing it seamlessly with his own growing anti-Semitism and anti-communist beliefs. In *Mein Kampf*, he blamed Germany's defeat on the noxious influence of international Jewry and Marxist elements. Nazi propaganda denigrated the democratic Weimar Republic it overthrew as an agent of betrayal, decrying it as "a morass of corruption, degeneracy, national humiliation, ruthless persecution of the honest 'national opposition'—fourteen years of rule by Jews, Marxists and 'cultural Bolsheviks.'" When Hitler took power in 1933, the *Dolchstoßlegende* became not just a fringe view but Nazi orthodoxy, taught as inerrant truth to schoolchildren and citizens alike. Jews especially were singled out for blame, branded disloyal elements who had betrayed Germany from the inside. This charge in turn became a license to dehumanize, and under Hitler the Nazi state rebranded Jewish citizens as parasites and traitors.

This myth-fueled dehumanization laid the groundwork for the most staggering and unfathomable deliberate destruction of innocent life in history. By the end of the Second World War in 1945, approximately 6 million Jews had been systematically executed by the Nazi state, and up to a further 11 million others had lost their lives, victims of what the Nazi machinery called the "Final

Solution"—what we now know as the Holocaust. Murder on this scale is simply impossible to comprehend, an ugly reminder of the human cost when sinister narratives take hold in a nation's psyche. We will never completely understand the bizarre mindset employed to justify such genocide, and we must be careful not to commit the reductive fallacy ourselves in trying to elicit answers to these horrifying questions. Still, it is fair to say reductive narratives played an ominous role in the callous scapegoating of Jews and others, reinforcing the prejudices of the perpetrators and collaborators.

Causal reduction fallacies come in multitudinous flavors, and perhaps the most pervasive of these are *false dilemmas* or false dichotomies. These assert a binary choice between extreme options, even when an entire ocean of options may exist. Despite their intrinsic hollowness, false dichotomies are supremely well suited to demagoguery, narrowing spectra of possibilities down to just two or so choices. If this inherently reductive rhetorical sleight of hand is accepted by an audience, the orator can readily present the outcomes as alternatively desirable or contemptible. Consequently, false dichotomies are inherently polarizing and not amenable to compromise. The Machiavellian trait of this fallacy is that it can be used to force the unaligned or nonpartisan to ally themselves with the speaker or lose face. It carries with it an implication that those not entirely in agreement with the proposal of the speaker are implicitly (or sometimes, incredibly explicitly) deemed the enemy.

This is nonsensical but surprisingly powerful, with a magnet-like ability to align the unwary in the direction the speaker wishes. Predictably, it has a long history of political deployment, most notably in the form of "you're either with us or against us" pronouncements, across all divisions of the political spectrum. Vladimir Lenin, speaking in 1920, declared: "It is with absolute frankness that we speak of this struggle of the proletariat; each man must choose between joining our side or the other side. Any attempt to avoid taking sides in this issue must end in fiasco."

Worlds apart politically, over eight decades later, President George W. Bush would use the exact same gambit in addressing a joint session of Congress in the wake of the 9/11 attacks, warning all nations listening that "Either you are with us, or you are with the terrorists." While Lenin and Bush would balk with contempt at each other's politics, there's a pleasing irony in the fact that neither had qualms about employing naked rhetorical falsehood to silence all but the most polarized views.

The long, ignoble pedigree of the false dilemma is impressive; historical examples would fill the rest of this book and volumes more besides. Arthur Miller's play *The Crucible* is set during the Salem witch trials, written in 1953 as a brilliant allegory for the overpowering hysteria of the then-prevailing anti-communist panic. In it, Deputy Governor Danforth invokes the fallacy, warning that "a person is either with this court or he must be counted against it, there is no road between." Outside politics, false dilemmas are used on emotive topics to push specific narratives, with logic often rendered unsound by the existence of other valid positions on a spectrum between the two extremes posited.

By their very nature, false dichotomies are antithetical to rational discourse, fostering extremism. The inherent polarization of a false dilemma can poison pragmatic solutions and dash constructive dialogue. Its deep intrinsic appeal lies in its ability to compress an entire spectrum down to simple, mutually opposed extremes, explaining its long-standing appeal to despots and demagogues. It is, however, rather telling that its corrosive influence has not reduced with time; it is still employed in a wide range of fields, with tedious predictability. Social media is rife with precisely this phenomenon, where complex topics with a wide scope for nuanced views get distilled down to a shouting match between two binary and diametrically opposed interpretations. In these forums, the spectrum of opinion becomes curiously bimodal.

The appeal of reductive fallacies is relatively easy to grasp: They offer simple, soothing explanations for complex phenomena. The illusion of understanding is reassuring and affirming, a psychological comfort blanket and totem of protection in a confusing world. The urge to understand cause and effect is something primal and intrinsic to the human condition–this enduring desire has been the engine that has driven mankind's development and intellectual appetite for millennia. It has led us to everything from taming fire to formulating quantum mechanics. Without this irrepressible drive to understand, we would be bereft of vast swaths of art and science. Yet, for at least as long as we've had the desire to understand, so too have we fallen victim to causal fallacies–it is written in the lingo of our superstitions, our rituals, and even our religions. As we will see in the next chapter, however, it can be remarkably difficult to separate cause from effect and far too easy to err, to our collective detriment.

5

SMOKE WITHOUT FIRE

Superstitious Pigeons, Anti-Vaccine
Activism, and the Post Hoc Fallacy

We humans have an abiding penchant for superstitions. No matter how rational we pride ourselves on being, there are few of us who don't perhaps feel a tingle of anxiety when walking under a ladder or a rising dread if a mirror shatters. Some of us even shun animals, places, or numbers we deem unlucky. Triskaidekaphobia, fear of the number 13, is so common that some hotels deliberately avoid a 13th floor or room. In our defense, superstition isn't solely a human trait—the great psychologist B. F. Skinner demonstrated that we share this quirk with another species: the humble pigeon.

Skinner's insight came from a now-classic conditioning experiment, where the birds were rewarded with treats at random intervals from a mechanical device. After a slew of random rewards, the inquisitive pigeons came to believe that some aspect of their behavior was triggering the gifting events, adopting a litany of rituals to encourage this. The pigeons had successfully been conditioned, engaging in complicated dances to curry favor with the capricious lord of snacks. These actions were elaborate and complex, repeated by the birds eager to be rewarded. Skinner observed that:

> One bird was conditioned to turn counter-clockwise about the cage, making two or three turns between reinforcements. Another repeatedly thrust its head into one of the upper corners of the cage. A third developed a "tossing" response, as if placing its head beneath an invisible bar and lifting it repeatedly. Two birds developed a pendulum motion of the head and body . . . Another bird was conditioned to make incomplete pecking or brushing movements directed toward but not touching the floor.

While Skinner has an impressive and long list of scientific findings to his credit, it's hard not to feel the zenith of anyone's career is "making pigeons superstitious." To Skinner, this was clear evidence that behavior is reinforced—when the pigeons performed their ritual, they were rewarded. The pigeons did not seem to question whether their system was robust; it just seemed to work. But we shouldn't pass too much judgment on the pigeons for this faux pas—after all, humans do it all the time. The intricate dances the birds performed were in many respects directly analogous to human rain dances, performed for centuries by tribes across America, Europe, and Asia. Rituals such as these are deeply woven into the fabric of our society, for good reason. We are keen observers, gifted with the ability to make inferences from observation—a trait that has long served us well.

But while the urge to connect two or more disparate phenomena is fundamentally human, the mere fact that one event preceded another is not in itself proof that the first event caused the second. It is frequently not easy to determine whether there is a causal relationship between the two observables, or whether the chronology is nothing more than a coincidental artifact. Yet the error of leaping to the conclusion that one event caused another based solely on this is ubiquitous. The class of informal fallacy that deals with cause-and-effect errors is included under the umbrella term *post hoc, ergo propter hoc* ("after this, therefore because of this"), capturing the essence of the error with charming brevity. Such questionable cause fallacies are superficially appealing, as they offer a sequence of events that seems integral to causality, but the vital credo is that a mere sequence does not come with any guarantee of cause and effect.

Take, for example, the scourge of malaria, an illness that has afflicted humans for millennia. In 400 BCE, long before the ailment had acquired its modern name, Hippocrates discussed its causation, asserting that malaria arose due to the unhealthy air in swampy environments. Given that Hippocrates is regarded as the father of medicine, and physicians to this day take an oath in his name, it is perhaps not surprising his influence cast a long shadow. Roman physicians too noted that those stricken tended to reside near marshes and swampland, with those who took walks in the night air disproportionally affected. This observation is at least a sensible one by the standard of medical antiquity. By way of contrast, contemporary physician Quintus Serenus Sammonicus beseeched patients to inscribe the word "abracadabra" on an amulet, and then write the word on paper several times, losing a letter upon each iteration as a cure for fever.

The link between dank environs and sickness endured, reaffirmed by generations of physicians. The connection is reflected in the name the illness eventually acquired: malaria, or "bad air." It wasn't until 1880 that French army surgeon Charles Louis Alphonse Laveran discovered parasites in the blood of a malaria victim. Just a few years later, in 1887, Ronald Ross, a British officer in the Indian Medical Service, demonstrated that mosquitoes could transmit the malaria parasite, in the process identifying the most persistent vector for the condition. And as mosquitoes are nocturnal and breed on stagnant water, the ancient connection between marshland at night and risk of illness was correct. Their inference, however, was bogus. It was not poor air that caused the illness, but parasites transmitted in the bite of mosquitoes, which happened to breed and feed near water.

Although the malaria and stagnant air link was misunderstood, the conclusion drawn was unlikely to cause any harm. If anything, it might even have saved lives inadvertently by keeping people away from areas where mosquitoes would feed and spread infection, much as Semmelweis's conclusions about hand-washing—for the wrong reasons—saved the lives of many young mothers. But this is happy accident, and the converse is every bit as frequent. Sometimes a conflation falsely made is not benign but riddled with severe consequences. Few incidents epitomize this better than the abject panic of the fabled vaccination–autism link of the late 1990s and early 2000s. Vaccination is, after clean water and sanitation, the single greatest life-saving measure on the planet. Despite this, there has been opposition to inoculation and vaccination since the beginning. A 1772 sermon by Reverend Edmund Massey was charmingly titled "The Dangerous and Sinful Practice of Inoculation," arguing that diseases were divine punishments from God; preventing smallpox was thus a "diabolical operation" tantamount to blasphemy.

Others opposed vaccination on the rather subjective grounds of bodily integrity or based on complete misunderstandings of how immunization works. Such opposition was often self-limiting; in 1873 in Stockholm, vaccine opposition rooted in religious views and apprehension about individual rights caused the Swedish capital to have a smallpox vaccination of a paltry 40 percent relative to the more serviceable 90 percent the rest of the country enjoyed. This somewhat stubborn position was quickly reversed following a massive outbreak of smallpox the following year, driving a marked uptick in vaccination as the epidemic reached a climax. In Stockholm at least, the visceral reality of smallpox shattered any lingering complacency citizens might have held.

In Europe of the 1800s, such an outbreak was no small ordeal—at the time, the disease claimed over 400,000 lives a year and rendered one-third of its survivors blind. Those stricken were covered in pus-filled sores and often left with permanent scarring. It paid no heed to rank or status, striking prince and pauper alike. Among countless victims who lost their lives to the deadly reign of smallpox were Queen Mary II of England, Emperor Joseph I of Austria, King Louis I of Spain, Tsar Peter II of Russia, Queen Ulrika Eleonora of Sweden, and King Louis XV of France.

By the dawn of the twentieth century, new insights into immunology gave rise to vaccinations for diseases that had plagued humankind from time immemorial. Smallpox sat atop this awful list, and by 1959 was killing upward of 2 million people a year. That year, a concerted worldwide effort to immunize against smallpox began in earnest. By 1979, the virus had been completely eradicated. For the first time in human history, a deadly virus had been relegated to the history books and bad memories, existing only in carefully controlled samples in a handful of biohazard laboratories across the world. Vaccinations against once-common illnesses like polio and measles were developed in the 1950s, saving countless more lives and consigning the misery of many diseases to distant memory. By 1994, the Americas were completely polio-free, followed by Europe in 2002.

But in many respects, vaccines were to become victims of their own success. The once-inescapable reminders of the potency of these illnesses began to fade from cultural consciousness. No longer did people encounter the pockmarked faces of smallpox victims, nor the crippling deformities of those who had endured polio. The haunting specter of children killed or rendered deaf or brain-damaged by measles infection slipped away from common shared experience. As the risk became more abstract and less obvious, complacency began to set in. People began to forget how profoundly vaccination had changed our world.* For most of the twentieth century, acceptance of vaccination was high, bar a persistent fringe that held immunization in contempt. This cohort remained bubbling in the background, attributing every malady conceivable to vaccination. For the most part, their assertions were outlandish enough to be ignored.† At the twilight of the twentieth century, young parents didn't worry about their children dying in infancy like their parents before. Survival was now a given, taken for granted—but some found new fears to consume them.

Chief among these was a growing concern about developmental disorders. In the late twentieth century, rates of autism in children had apparently begun to rise, terrifying parents. The hallmarks of autistic spectrum disorders tended to manifest not long after

---

* During the First World War, poet and soldier Siegfried Sassoon decried the "callous complacency with which the majority of those at home regard the contrivance of agonies which they do not share, and which they have not sufficient imagination to realize." Sassoon was referring to the blind indifference of the public to slaughter out of sight, but I often think of this sentiment when I encounter those dismissive of vaccines.

† Psychological traits of anti-vaccine activists include reasoning flaws, reliance on anecdote over data, and low cognitive complexity in thinking patterns. Conspiratorial thinking is endemic, with critics derided as agents of some malign interest group.

toddlers had been given their immunization. For some, this suggested an uncomfortable implication—perhaps immunization itself induced autism? There was, however, no medical evidence supporting this grim sequence, and plenty contradicting it. This spurious link might have slowly ebbed from the periphery of the public consciousness had it not been for the actions of an infamous British gastroenterologist, Andrew Wakefield.

In 1998, Wakefield and coauthors published a small study involving 12 autistic children in the respected medical journal *The Lancet*, claiming to have discovered a constellation of intestinal symptoms associated with autism. They christened it *autistic enterocolitis*, and buried deep within the paper's discussion section was the rather speculative suggestion that this might perhaps be related to the measles vaccination. It existed as a fleeting, tentative assertion, bereft of any corroborating information. Under normal circumstances, this flimsy conjecture would have been dismissed as baseless, had Wakefield not done something rather unusual—he called a press conference. Unshackled by the constraints of diligent scientific conduct, Wakefield announced that he'd uncovered evidence linking the Measles-Mumps-Rubella (MMR) vaccine with autism, and that the triple vaccine was unsafe—a claim that jibed with the growing undercurrent of concern about the increased prevalence of the disorder.

Initially at least, Wakefield's startling claims had little effect on public discourse. His assertions were contradicted by a mass of much stronger data. Mainstream science and health journalists were savvy enough to spot the hallmarks of dubious science and wary of the self-promotion Wakefield displayed in spades. Nevertheless, slowly a dedicated core of vaccine opponents propelled the story into the mainstream, evading the gatekeepers of science journalism by pitching it as a human-interest story to more credulous journalists. Nonspecialist journalists presented themselves as voices of concern, emphasizing that autistic traits manifested after vaccination, cementing the impression of cause and effect. For anti-vaccine activists who needed public panic to push their case, this was a godsend; by

2002, approximately 10 percent of all published science stories in the UK concerned the MMR vaccine, with 80 percent of these stories penned by journalists with no grounding in science or medicine. Physician and author Ben Goldacre summarized the absurdity of the situation concisely: "Suddenly we were getting comment and advice on complex matters of immunology and epidemiology from people who would more usually have been telling us about a funny thing that happened with the au pair on the way to a dinner party."

This glaring disconnect between expertise and coverage should have been telling; science journalists were by and large familiar with the long-standing tendency of anti-vaccine campaigners to misinterpret clinical evidence, and well-versed enough in the scientific method to dismiss Wakefield's claims. When dedicated science writers reported on MMR, they tended to emphasize that the strength of evidence for the benefits of vaccination was overwhelming while the evidence for an autism link was essentially nonexistent. In all the baying about protecting children, reminders from the scientific community that vaccination saves lives on a huge scale went sorely unheeded in the maelstrom of panic. As the saying goes, paper never refused ink, and the scientific ineptitude of the journalists, celebrities, and public figures engaged in such contemptible fearmongering was completely overlooked. Media outlets began fawning on Wakefield and his supporters; the *Telegraph* even lionized him as a "champion of patients"—roundly ignoring medical consensus on the safety and efficacy of immunization.

Inevitably this uproar came at a high cost. Within months, vaccination rates across Western Europe plummeted. This was an incredibly dangerous situation. Measles is especially infectious, being airborne, readily transferable, and difficult to avoid. Each single case leads to an average of 12 to 18 secondary infections. Not only is infection deeply unpleasant, but its side effects can be devastating, including hearing loss and brain damage. It can also be deadly, killing over 160,000 annually. The measles vaccination saves over a million lives a year, yet there is no room for complacency.

Due to its tenacity, a high collective resistance is required to prevent it from becoming endemic. Individuals with immunity provide
a "firewall" that protects those who cannot be vaccinated—young
babies and those with medical conditions that rule out immunization. For a disease as virulent as measles, this herd immunity must
be around 94 percent to prevent outbreaks.

Thanks to the credulous and frankly deplorable conduct of many
press outlets, Wakefield's dubious message spread far and wide.
The tentacles of Wakefield's panic even worked their way across
the Atlantic Ocean, as we'll see later. The net result was an entirely
exaggerated picture of vaccine risk,* despite the overwhelming abundance of evidence showing the safety and incredible effectiveness
of immunization. Across the UK, the epicenter of the controversy,
immunization rates dived as low as 62 percent. Outbreaks of the
incredibly virulent disease went from a relative rarity to an all-too-
common problem. Across the Irish Sea in Dublin, low vaccination
rates provided the ideal circumstances for the virus to run rampant,
with three children dying and several others permanently scarred.

One honorable exception to the appalling standard of conduct
shown by news organizations was investigative journalist Brian
Deer. Deer turned a skeptical eye to the increasingly shrill cries
over vaccination, cognizant of the fact that Wakefield's claims
were directly opposed to the bulk of scientific data. In 2004, Deer
published evidence that Wakefield had received £55,000 (about
$75,000) in payments from lawyers seeking evidence to use against
vaccine manufacturers. In defiance of scientific ethics, Wakefield
had not declared this potential conflict of interest. Deer also presented damning evidence that Wakefield had applied for patents
on a rival vaccine to MMR, and that he was fully aware of results
from his own laboratory that stood in direct contradiction to the
claims he had made in public. *The Lancet* conceded that Wakefield's
work was "fatally flawed"; Wakefield responded by suing Deer for

---

* This is an archetypal example of "false balance," a topic we'll address in part 5

libel, a transparent and heavy-handed bid to stem a mounting tide of evidence against him. Fortunately for the world, this severely underestimated Deer's tenacity, who continued to publish more damning findings on Wakefield's conduct. By 2006, Deer had revealed not only that Wakefield's claims were completely without merit but that he had also received the sum of £465,653 (about $620,000) from trial lawyers who expected him to find evidence that the MMR vaccine was damaging. Ultimately Wakefield was forced to drop the case and pay all costs.

This proved the death knell for the acceptable cult of Wakefield. The UK's General Medical Council began a thorough investigation, and *The Lancet* retracted his papers after finding evidence of scientific fraud. In April 2010, Deer demonstrated that Wakefield had doctored evidence. A month later, the GMC panel found Wakefield guilty of serious professional misconduct involving professional dishonesty and the abuse of developmentally challenged children. Wakefield was struck off the register and Deer published evidence that Wakefield had planned to sell medical tests for the fictitious condition, which would have netted him an estimated $43 million a year. Wakefield's fall from media darling to medical pariah was complete. Professor Fiona Godlee, editor of the *British Medical Journal*, summed up the despicable nature of Wakefield's conduct, pulling no punches in her assessment:

> Who perpetrated this fraud? There is no doubt that it was Wakefield ... [He] has been given ample opportunity either to replicate the paper's findings, or to say he was mistaken. He has declined to do either. He refused to join 10 of his co-authors in retracting the paper's interpretation in 2004, and has repeatedly denied doing anything wrong at all. Instead, although now disgraced and stripped of his clinical and academic credentials, he continues to push his views. Meanwhile the damage to public health continues, fuelled by unbalanced media reporting and an ineffective response from government, researchers, journals and the medical profession.

Autistic enterocolitis was a myth, irreproducible and underpinned only by evidence Wakefield had faked.* Yet, even with all this evidence marshaled toward an inescapable conclusion of fraud, many still rally around Wakefield, convinced their child's autism must have been linked to MMR. Their strongest evidence for this belief is that, at some arbitrary time after vaccination, their child began to exhibit signs of autism. This is the post hoc, ergo propter hoc fallacy at its terrible extreme—while offering an appealing simplicity, the conclusion simply does not follow. Rising rates of autism had nothing to do with vaccination, with the most likely culprit being widening diagnostic criteria for autism. Nor was the appearance of autism after vaccination particularly surprising; autism manifests in early childhood and the telltale markers such as impaired communication tend to become apparent by age two or three, not long after immunization. The erroneous attributing of cause and effect was enough to lend false credence to a manufactured panic.

While peak paranoia over MMR might have subsided by the early 2000s, the child victims of that era were not the only ones to suffer. Parents apprehensive about vaccination refused to get their children protected, and this fear diffused slowly across the world. Those toddlers grew without immunization or the requisite background levels for herd immunity, with predictable results across both Europe and America. In 2011, there were over 26,000 cases of measles in Europe, with nine deaths and 7,288 hospitalizations. By 2018, that figure had climbed to over 82,596. In 2012, the UK saw cases

* It is worth noting that, initially at least, Wakefield focused his ire on the MMR vaccine specifically, advocating single-dose vaccines for which he had surreptitiously applied for patents. Accordingly, some journalists didn't believe Wakefield to be anti-vaccine. This was misguided, for the genesis was an anti-vaccine canard that "vaccine-overload" damages children, a known falsehood even then. It's also worth noting that anti-vaccine activists often avoid referring to themselves as such, instead using euphemisms such as "pro-safe vaccines." But this is just a semantic game—if one amplifies anti-vaccine claims while completely ignoring the overwhelming evidence of safety and efficacy, one can hardly be said to be acting in good faith. If you only question scientific evidence but give any old anecdote a free pass, then you're evidently stacking the deck. One's actions define whether one is anti-vaccine (or racist or misogynist, etc.) or not, and in this respect self-identification is pretty much irrelevant. Wakefield's anti-MMR stance was an ostensibly respectable vehicle to project tired old anti-vaccine nonsense. Predictably perhaps, his later years saw him abandon any pretence of being anything other than anti-vaccine.

surging to a 20-year high, with a single outbreak in Wales in 2013 infecting 1,200 people and killing one. Ireland saw 443 cases in 2010, more than double the previous year. In North Cork, vaccination rates as low as 26.6 percent were recorded.

America—once practically measles-free—has seen an endemic rise in infection rates. Tarnished as Wakefield's halo might be, his toxic legacy remains—in the last decade, vaccination rates across America have steadily plummeted, driven by anti-vaccine propaganda online, much of it centered around Wakefield's bogus claims.* In 2014, there were 677 cases in 27 states, a 20-year high. A single infected person led to at least 150 cases of the disease in Disneyland the following year, with authorities noting that "substandard vaccination compliance is likely to blame for the 2015 measles outbreak." Early 2019 saw New York endure its worst measles outbreak in decades, contributing in part to 2019's setting records for measles outbreaks worldwide. These victims are a legacy of the vaccine panic, a fear still promoted by dedicated anti-vaccine activists. The WHO wearily notes that this itself is nothing new: "How one addresses the anti-vaccine movement has been a problem since the time of Jenner.† The best way in the long term is to refute wrong allegations at the earliest opportunity by providing scientifically valid data. This is easier said than done, because the adversary in this game plays according to rules that are not generally those of science."

Such is the extent of the problem that in 2019, for the first time, the WHO declared vaccine hesitancy a top-ten threat to global health. The MMR panic itself was supported by the mere observation that vaccination sometimes preceded the manifestation of

---

* Not that this has been any impediment to him—at the time of writing, Wakefield lives in the United States, where he is celebrated as a hero by the anti-vaccine community with profitable speaking engagements. He is also in a relationship with supermodel Elle Macpherson, and was lauded by Donald Trump after his election to president. Truly, the continuing cult of Wakefield is perhaps a cynic's retort to the very concept of karmic justice.

† Edward Jenner invented the first vaccination in 1796.

autism. This coincidental timing functioned as a Trojan horse for dedicated anti-vaccine activists to push the post hoc, ergo propter hoc fallacy to the naive, fueling the entire destructive rush. The consequences we feel today are a telling reminder of the consequences of faulty thinking. The panic also had another element that fanned those flames—the cultural zeitgeist of the time. Looking back at the whole affair, one might wonder why so much angst was expended over a hypothetical risk of autism and why that fear resonated more with people than the reality of what vaccination was preventing.

Part of the answer lies in availability—for the parents of the early 2000s, images and stories of the children dead or permanently impaired by measles weren't in our cultural lexicon. Due to research and public health effort years before, the graphic consequences of the virus were no longer frequent occurrences and so lacked any resonance with concerned parents. Autism, by contrast, was so frequently discussed as to be part of common vernacular. Magazines and newspapers constantly ran stories on the challenges facing autistic children, speculating about the causes behind the apparent rise in rates. These speculations tended to ignore the mundane but important fact that the diagnostic criteria for the condition had widened substantially in previous years, and that children who might formerly have been classified as intellectually disabled were finally being recognized as being on the spectrum. Children who had formerly been institutionalized and effectively invisible were suddenly in the public eye. The idea of "autism" was available to the public mind; the devastating impact of measles was not. And this conceptual "availability" skewed perceptions deeply—and tragically.

In 2018, a terrible mistake in Samoa saw two children die while getting their vaccines, when the vaccine powder was accidentally mixed with expired anaesthetic instead of diluent. Anti-vaccine activists capitalized on the understandable fear this caused, and by 2019, vaccination had plummeted to 31 percent. In August that year, a measles-infected passenger arrived in Upolu; by early 2020, there were over 5,700 cases in a country of just over 200,000 people, with 83 deaths. Even while the government of Samoa instituted a rapid vaccine campaign that saw 94 percent of the population vaccinated by December, anti-vaccine activists tried to undermine it. But with over 3 percent of the entire population directly affected, the falsehoods anti-vaccine activists relied on rang increasingly hollow.

This phenomenon of affording more weight to easily accessed or recent information is known as the *availability heuristic*. When evaluating a concept or forming an opinion, this is in effect a mental shortcut that relies on immediate examples that are easy to recall. This pivots on the assumption that if something is easy to recall, it is therefore important—or, at least, it's more important than alternative explanations. The easier it is to recall information, the greater stock we place in it. And in effect, this usually biases our opinions toward recent information or memorable examples. But the mere fact that some information is recent or memorable doesn't make it true, nor are any conclusions drawn from this shortcut watertight. It was easier for concerned parents to access scare stories about autism rather than measles deaths. Similarly, in the case of Samoa, the tragic deaths of two children due to a mix-up held a visceral power, until it was superseded by the awful reality of what low vaccination rates truly entailed.

The availability bias is but one type of mental shortcut in an entire family of heuristics. Sometimes the reason for this is one of speed over quality. When it comes to survival, for instance, there are advantages to rapidity. Suppose we're out in the wilderness and suddenly we encounter a rustling in the undergrowth. Most likely, it's nothing harmful; perhaps it's the wind or birds or a fox. Depending on where we are and what we know, we could probably calculate the most likely proximate causes. But in general, that's not what we do—our mind jumps to react, our sense of danger primed. This course of action might be life-saving if the noise is not something benign but a snake lurking beneath.

These decisions and our reactions are so fast they seem to evade active thought itself. These rules of thumb are heuristics, striving to keep us alive by short-cutting our reasoning to err on the side of caution. They're far from perfect, of course, but they function as an autopilot. Psychologist Daniel Kahneman classified our thinking into two distinct states: system 1 and system 2. In Kahneman's framework, system 1 is our fast, intuitive, and seemingly automatic response; by contrast, system 2 is the slower, more analytical mode of thought where reasoning dominates. These systems are complementary; logical thought is cognitively expensive, while heuristics keep us alive. In Kahneman's words, "This is the essence of intuitive heuristics: When faced with a difficult decision, we often answer an easier one instead, usually without noticing the substitution."

Heuristics do much more than protect us from snakes—they're fundamental to how we think. Even when thinking analytically, we employ elements of both systems—so much that heuristics are embedded in our reasoning. Kahneman and his colleagues have identified a trove of them that sit at the very heart of our ability to make inferences. The problem is that inferring from these easily recalled examples is fraught with the potential for serious error. Usually, the examples we recall most strongly are the most emotionally charged rather than the most representative cases. People dramatically overestimate their risk of death from terrorism and violence, while completely underestimating much more likely killers such as heart disease and stroke. But clearly, when we rely solely on heuristics, we run a risk of inferring incorrectly and ultimately erring in our reasoning. In Kahneman's words, "Heuristics are quite useful, but sometimes lead to severe and systematic errors."

Intuitive heuristic reasoning might be fast but it's fraught with pitfalls. Take the simple question posed by Kahneman in his 2011 book *Thinking, Fast and Slow*. A baseball and a baseball bat together cost $110—the bat costs $100 more than the ball. How much does the ball cost? Intuitively, most people spring to the answer of $10. But that answer is wrong, for then the bat itself would cost $110 and the total cost would be $120. The actual answer is found by translating the word problem to algebra. If x is the cost of the bat and y the cost of the ball, we have two simple equations:

$$x + y = 110$$
$$x - y = 100$$

This is a simultaneous equation, and adding both equations, we arrive at $2x = 210$. This means x, the baseball bat, costs $105. From this, we find that the baseball costs $5. If you got it wrong, don't feel bad—as Kahneman explains, this error is ubiquitous in even the brightest:

> Many thousands of university students have answered the bat-and-ball puzzle, and the results are shocking. More than 50 percent of students at Harvard, MIT, and Princeton gave the intuitive—incorrect—answer. At less selective universities, the rate of demonstrable failure to check was in excess of 80 percent . . . many people are overconfident, prone to place too much faith in their intuitions. They apparently find cognitive effort at least mildly unpleasant and avoid it as much as possible.

We talk of relying on our gut or instincts, but our automatic reactions are generally suboptimal for anything requiring a modicum of nuance or analysis. The jarring reality is that, when confronted with a decision to make, our implicit reliance on what is fast and feels right can often throw us askew, with sometimes dangerous consequences. We must be incredibly wary when inferring cause and effect from limited data, and cautious of jumping to unjustified conclusions. In defiance of the tired old adage, smoke frequently occurs without fire. The tragedy is that, in searching for phantom fires, we often end up kindling our own infernos.

6

THE NATURE OF THE BEAST

Misguided Essentialism, the Illogical Foundations of Racism, and the Trial of Galileo

Few issues are as likely to expose such naked tensions as immigration. Worldwide, there is a tangible fear in some quarters of being overrun by alien invaders. This purported invader need not even be particularly alien—these fears can extend to people in the same nation, divided along ethnic or racial lines. The United States, with its complicated racial history, is perhaps the prime example of this. Slavery was the fuse that ignited the flames of the American Civil War. And while Abraham Lincoln's victory may have emancipated 4 million Black slaves, even after liberation they were often ostracized and discriminated against, relegated to the fringes of society. Long after the war, racial segregation was enforced alongside voter disenfranchisement, confining Black Americans to perpetual second-class citizenship, with frequently violent oppression.

The genesis of the civil rights movement in the 1950s gave a glimmer of hope for an improved world. On August 28, 1963, Martin Luther King Jr. addressed a crowd of 250,000 civil rights supporters at the foot of the Lincoln Memorial, hopeful for the day that people "will not be judged by the color of their skin but by the content of their character." Before the decade had ended, the Voting Rights Act and Civil Rights Act had been passed, making racial discrimination or abuse federal crimes. Sadly, Dr. King did not live to see this, having been assassinated in 1968. By the early 1970s, newspapers were openly wondering if there might soon be the dawn of a true "post-racial" society, the culmination of King's dream, where race no longer had any bearing on one's destiny, and where prejudice based on one's melanin level was no more than a relic of a backward time. The election of Barack Obama to the office of president in 2008 brought heady optimism that the era of the post-racial society might finally have arrived.

Alas, this was more wishful thinking than anything else. Old attitudes might have been suppressed by legal obligation, but they had not gone away—they had simply been better concealed. While in office, Obama endured constant racism, both overt and subtle. A conspiracy theory quickly emerged that he had been born in Kenya and thus his presidency was illegitimate. This so-called "birther" movement claimed that his birth certificate was fake and that he was secretly Muslim—a none-too-subtle attempt to discredit him and render him "un-American." To Obama's credit, he handled such accusations with characteristic grace and panache. But fundamentally, there were those who couldn't accept this successful Black man was a "true" American. The loudest of all the "birthers" was undoubtedly reality TV star Donald J. Trump.

When Trump ran for president in 2016, he articulated and reiterated a strain of naked racism and xenophobia. He vilified immigrants, making the building of a quixotic wall on the US-Mexico border a central pillar of his campaign. The central message was that immigrants and outsiders were the cause of the country's woes, and Trump promised to stamp them out. In a progressive world, such remarks should have sunk any candidacy; instead his open xenophobia appealed to white nationalists, who crawled out of the woodwork, emboldened to be openly racist. David Duke, former leader of the Ku Klux Klan, endorsed Trump, enthusing that "he's going to take our country back." Trump's campaign was led by Steve Bannon, whose website, Breitbart, made White nationalism a central theme. White supremacist Richard Spencer supported Trump, and the nascent "alt-right" movement rallied behind him. Neo-Nazi groups and White nationalists across the country praised him openly. Their embrace should by all rights have been the kiss of death, but in November 2016 Trump won the election—to the elation of White supremacists and the despair of so many others.

Around the same time, across the Atlantic, things were no less heated. In the run-up to the 2016 Brexit referendum in the UK, reported incidences of hate crimes soared. Negative views concerning immigration were identified as the single biggest predictive factor for "Leave" voters. Campaigning often featured xenophobia. Nigel Farage, demagogic leader of the UK Independence Party, unveiled an anti-migrant poster showing a line of dark-skinned people waiting to cross a border. This was depicted to be a consequence of continued EU membership, despite this being a racist fiction divorced from reality. White nationalism was frequently invoked, with charged rhetoric giving way to violence. Labour politician Jo Cox was gunned down in the street and stabbed, her assailant chanting nationalistic slogans. At his trial, his twisted justification for the brutal murder of an inspiring and innocent young woman was that Cox's support for immigrants and the European Union marked her as one of "the collaborators," "a traitor" to white people.

White nationalism isn't a new phenomenon, alas, and is endemic across America, Europe, and Russia. The unifying view is that white people are a race, with a common cultural or ethnic identity that needs to be preserved. These individuals believe that multiculturalism, low White birth rates and non-White immigrants pose a threat. Others go further, believing that racial integration is a ploy to overrun majority White countries, a process they label "White genocide." Supremacist views are frequently aired, asserting that whites are better than other races in terms of intelligence, art, heritage, or other traits. In these narratives, White people are depicted as a culture under siege from fundamentally different outsiders. These unfortunately can't be dismissed as mere fringe views. In late 2017, the University of Virginia's Center for Politics carried out a US-wide survey gauging attitudes to racial tension. The results made for uneasy reading: 31 percent of respondents agreed with the statement that America needs to "protect and preserve its White European heritage," while 39 percent agreed with the claim "White people are currently under attack in this country." The year 2020 has seen the specter of racism and White supremacy increasingly dominate media headlines, as the resurgence of the Black Lives Matter movement responded to yet more killings of Black Americans at the hands of police—much of this encouraged or endorsed by Trump himself.

Like most powerful myths, immigration fears hold a grain of truth, albeit one that has been warped utterly. It is true that birth rates in Europe have fallen, but this decrease in fertility is not surprising, and nor is it a sign of "White oppression." Across the globe, high female literacy is directly associated with contraceptive use and lower birth rates. Educational attainment also decreases the number of children a woman is likely to bear, defined as the Total Fertility Rate (TFR). This effect can be remarkably stark and is difficult to overstate. In Ghana, women who complete high school have a TFR of two or three, relative to a TFR of six for their contemporaries without an education. So, what of the concern expressed by many opponents of immigration that an influx of poorly educated foreigners will overwhelm natives?

Fears that immigrant populations will overrun indigenous people are not new—similar panic was voiced in 1860s America, where the higher birth rate of the immigrants—including among White immigrants like the Irish and Italians—raised concerns that foreigners would rapidly outbreed those already settled. The astute reader might detect a clear irony in White Americans of the 1800s (or indeed now) being concerned about immigrants, given the history of how the United States came to be. In any case, this alarm was ill-founded, as by the second generation, immigrant birth rates had already reduced substantially, tending toward the norm. This isn't surprising, as socioeconomic and educational factors have far more bearing on reproductive rates than any intrinsic virility.

But what is curious is that documentary sources from the 1800s attest that the Irish were considered a distinct race—a notion we'd consider daft today. This curious demarcation raises a loaded question: What precisely is race, and of what is it predictive? With a staggering amount of blood spilled for this concept over centuries, one might be forgiven for thinking race must have some measurable, objective basis. But surprisingly perhaps, race is not meaningful from a scientific perspective. The entirety of human life on the planet is one species: Homo sapiens. Genetically, humans differ by only minimal amounts—on average in a DNA sequence, each human is over 99.9 percent similar to another human. In the words of scientist Michael Yudell, "Genetic methods do not support the classification of humans into discrete races."* The belief that there are "essential" genetic or intrinsic traits that define ethnic groups has no real scientific basis. Indeed, there's abundant evidence that variation within ethnic groups far outstrips those between groups. Genetic characteristics might be associated with certain populations, but these characteristics are not in any way exclusive to a population. Lines of demarcation are entirely arbitrary, unsupported by evidence. From a scientific perspective, "race" is so nebulous a term as to be useless. To understand why, we simply need to look at the one thing that unites White supremacists worldwide: their skin color.

White skin is the archetypal European or Aryan trait. Yet this relatively simple mutation is quite recent, and complex in origin. The first modern humans to settle Europe from Africa about 40,000 years ago had dark skin, a distinct advantage in latitudes with ample sunlight. Dark skin was the norm across central Europe 8,500 years ago. In the far north of the continent, natural selection had begun to select for lighter skin. Bodies recovered from a 7,700-year-old archaeological site in Motala, Sweden, had the genes SLC24A5 and SLC45A2, which give rise to depigmentation and lighter skin, and the HERC2/OCA2 gene, thought to give rise to blue eyes and lighter hair. These mutations were advantageous in conditions of low light, maximizing synthesis of vitamin D. Our ability to digest cow's milk also evolved as a strategy to maximize vitamin D around the same time.

* Adam Rutherford's *A Brief History of Everyone Who Ever Lived* gives great insight into our shared human heritage.

For centuries, white skin in Europe was confined to the north-ernmost part of the continent. That divide remained until the arrival of the first farmers from the Near East, who carried genes for both light and dark skin. They bred successfully with the resident hunter-gatherers so that, over time, lighter skin became more common across Europe. The gene variant SLC24A5, once rare, exploded in frequency across the continent only around 5,800 years ago. Far from being an essentialist trait, the emergence of white skin as a common phenotype required the sustained interbreeding of many different groups. This makes the spectacle of White supremacists touting their racial purity as deluded as it is disgusting.

The reality is that there is precious little difference between us. Fear of being bred out of existence displays an ignorance of the reality that all humans alive today are closely related, with only tiny genetic variations across our entire species. If anything, increasing our genetic diversity is beneficial, and avoids the dominance of deleterious recessive traits. Cystic fibrosis (CF), for example, may be passed on if both parents carry a mutated gene copy. In Ireland's island population, this gene mutation is carried by 1 in 19 people. Consequently, Ireland has the world's highest rate of CF. Exclusive in-group breeding is detrimental to survival, circumvented only by population diversity. We have always been a curious, promiscuous species—a trait that has always stood to our benefit. The "White race" is for all intents and purposes a fiction, and White supremacy just a cover for pathetic individuals to claim tenuous links to some questionable past glories.

I should take a moment here to address a frequently derailing criticism. One might object that, if race is so meaningless a concept, how might we explain the apparent differences in intelligence between ethnic groups? In the US, much has been written on the seemingly superior intelligence of Ashkenazi Jews, who tend to score high on IQ tests. Black Americans reportedly score lower on IQ tests than whites. But there's a massive confounder here: IQ test results are hugely dependent on social and educational factors. Adequate childhood nutrition is especially important—iodine deficiency reduces IQ by an average of about 12 points. Social factors and parental educational attainment also influence IQ scores. In America, Black families are still more likely to be malnourished and with lower levels of education than more affluent Whites. These discrepancies have nothing to do with poor choices on the part of Black Americans or virtuous living by Whites, but a function of endemic poverty in which many Black people still remain, the human consequence of decades of enduring oppression.

The racial IQ gap has also been closing for decades, far too quickly to be due to genetics. Kenya provides a striking example: between 1984 and 1998, IQ rose by 26.3 points. This increase reflected the fact that national nutrition, health, and parental literacy had improved. Nor is the superior performance of Jews on IQ tests an artifact of genetics—during the First World War, IQ tests conducted on Jewish soldiers yielded lackluster results. Analyzing the data, psychologist and eugenicist Carl Brigham stated that this "would rather tend to disprove the popular belief that the Jew is highly intelligent." By the Second World War, only decades later, Jewish IQ scores were above average. This would not have surprised Alfred Binet, co-inventor of the IQ test, which was created in France to identify struggling students so that they could be given extra support. From the outset, Binet had stressed that intelligence was diverse and that intellectual development progressed at variable rates and with substantial environmental influence. His stance was that intelligence was malleable, not fixed. This clearly wasn't heeded by some, and the metamorphosis of these tests from noble intervention to discriminatory metric must have caused Binet no end of despair.*

Racist debacles expose a persistent pitfall in human thinking, tapping into a philosophical debate that has raged for millennia:

---

* To be clear: Although IQ tests can be abused, they are important tools that yield useful measurements. *Intelligence: All That Matters* by Stuart Ritchie is a great primer.

the question of the essential nature of things. *Essentialism* is difficult to adequately outline without delving into the rich history of philosophical thought, but a simple working definition is that for any given thing, concept, or group, there exists a set of attributes essential to that identity. This idea is an ancient one—Plato's dictum that all things have some essential perfect form behind them is now known as Platonic idealism. Aristotle too thought along similar lines, which linguist George Lakoff concisely summarizes as "those properties that make the thing what it is, and without which it would be not that kind of thing."

Essentialism has ample merit for many applications—mathematics being a prime example. Here, definitions are of utmost importance and the properties of sets must be clearly defined. Mathematician Gerald Folland stated that "it is a truth universally acknowledged that almost all mathematicians are Platonists, at least when they are actually doing mathematics." Yet caution must be observed when applying similar reasoning outside such well-defined boundaries; if no intrinsic "essence" or nature can be discerned for a given group, then this approach is inherently doomed to failure or tragedy. In lieu of objective ways to determine these defining traits for many real groups, an unfortunate trend exists of people simply asserting that certain traits are in the essence of whatever group they refer to.

As we have seen, much of the underlying rationale for racism pivots on essentialism—that races have intrinsic superior or inferior properties. But on closer examination, these asserted properties simply don't exist or are so vague and widely shared as to be meaningless. Not that any of this has stemmed the horrors of racial discrimination. While the philosophical debates surrounding essentialism are fascinating, we'll limit ourselves to exploring informal fallacies pertaining to some dubious trait. To circumvent confusion, we'll categorize this class of informals as *arguments from nature*. But it's important to note here that "nature" itself is a somewhat malleable term, allowing all manner of moving goalposts to be spirited around under its cloak.

Philosopher Anthony Flew envisioned a now classic example of this type of equivocal reasoning:

> Imagine Hamish McDonald, a Scotsman, sitting down with his *Glasgow Morning Herald* and seeing an article about how the "Brighton Sex Maniac Strikes Again." Hamish is shocked and declares that "No Scotsman would do such a thing." The next day he sits down to read his *Glasgow Morning Herald* again; and, this time, finds an article about an Aberdeen (Scotland) man whose brutal actions make the Brighton sex maniac seem almost gentlemanly. This fact shows that Hamish was wrong in his opinion, but is he going to admit this? Not likely. This time he says: "No true Scotsman would do such a thing."

This is what has come to be known as the *No True Scotsman* (NTS) fallacy.* Of course, the mere fact that one was born in Scotland (or anywhere, for that matter) does not preclude one from being a sexual predator, yet the fictional Hamish McDonald here implicitly assumes there to be some common characteristic set of traits inherent to being Scottish, including the tenet that a Scottish man cannot be a rapist. Instead of simply correcting his false assumption, Hamish clings to his wrong-headed definition by disavowing the wayward example.

While this example is entirely hypothetical, the NTS fallacy is often used to appeal to some notion of purity in a given group and a method of dismissing relevant criticism of bad behavior. Let's see what happens when we replace a couple of terms. Maybe "no true Scotsman" could be considered "un-Scottish." Then, could "no true American" perhaps be called "un-American"? Here we move from farce to a very serious, and very real, political example. In the 1940s, the infamous House Un-American Activities Committee

---

* Not to be confused with "True Scotsman," a term referring to the wearing of a traditional kilt without underwear. This is actually standard military uniform. The terms "going regimental" or "going commando" are frequently employed euphemistically to refer to an absence of underwear, so it's probably worth not muddying these two concepts.

performed an inquisition to sniff out (frequently phantom) scents of communism in American public life. Yet their moniker contains a glaringly obvious example of the NTS fallacy: America has long been a populous and diverse country, and there is no explicit contradiction in one being American and being interested in communism. Nevertheless, HUAC persecuted anyone deemed tainted by communist influence, including an entertainment blacklist that affected such luminaries as Charlie Chaplin, Orson Welles, Pete Seeger, and Harry Belafonte. Fittingly, in 1959, the committee was denounced by former President Harry S. Truman as the "most un-American thing in the country today."

In more recent years, the un-American label has been applied to everything from fees for plastic bags to collective bargaining, rendering them similarly flawed appeals to fallacy. As with all informal fallacies, nuance and context is incredibly important in determining whether dubious reasoning is being deployed. NTS arises when the groups involved are invoking some nebulous set of characteristics that are not necessarily vital to membership of that group. Yet if the characteristics in question are integral to membership and can be objectively defined, then it might be extremely relevant. For example, imagine that Hamish claims to be a pacifist, opposed to all forms of violence. As the pacifism movement was started in Glasgow, Scotland, in 1901, making Hamish a pacifist in this example seems rather fitting. Yet if he tears down Buchanan Street beating people senseless, it would certainly not be fallacious to note that Hamish is no true pacifist, given that his behavior is plainly at odds with the explicit position he ostensibly subscribes to.

Intimately related to this school of reasoning is the fallacy of *appeal to nature*. In such rhetorical tactics, a thing is asserted to be inherently good because it is "natural" or bad because it is "unnatural." Such reasoning is often deployed in the realms of alternative medicine, where purveyors of dubious products hawk their wares with the confident claim that their products are "natural," as if that inherently makes them better than conventional medical treatments. The galling lack of evidence for efficacy of these treatments aside, even the proud declaration of "natural" is intrinsically hollow. Even ignoring the slippery definition of what is and isn't natural, the logic underpinning this claim is deeply confused. We might perhaps charitably define "natural" as that which occurs without human intervention—but even under this lax definition, plenty of naturally occurring things can kill or maim us, from deadly nightshade to Ebola. Uranium and arsenic are "natural" but you would be ill-advised to sprinkle them on your breakfast cereal. The simplistic conflation of natural with healthy or good is a non sequitur, fatally undermined by the equivocal adjective "natural."

The same argument applies when the term "unnatural" is evoked. For instance, from the perspective of the Catholic Church, homosexuality is considered a deeply unnatural state of being, referred to in solemn Latin as *peccatum contra naturam*. Yet this too is simply an archetypical example of the fallacy of appeal to nature, which fails spectacularly with even a cursory glance at the natural world. Homosexual behavior is ubiquitous in the animal kingdom and has been documented in well over 1,500 species, from giraffes to elephants to dolphins and our primate cousins. And while much of this activity is nonexclusive and does not preclude dalliances with the opposite sex, exclusively homosexual pairings do occur. To cite one common example, about 8 percent of domestic rams will only form partnerships with other males and cannot be enticed into mating with ewes. The proclivities of the natural world are entirely relevant to us because, despite the lofty status we afford ourselves, we too are part of the animal kingdom, differing only in that we have a pre-frontal cortex evolved enough to give us the metacognition to be aware of that fact.

The most destructive variation of the argument from nature is the ever-popular *argumentum ad hominem* ("arguing against the man"). An *ad hominem* argument is essentially a personal attack, aimed at the speaker or the speaker's credibility rather than addressing the argument they posit. If the attack has no relevance to the speaker's argument, then the tactic is completely empty. It comes in several distinct flavors, the most common being the abusive or demeaning form. Examples in rhetoric are abundant, especially in political arenas where character-smearing and undermining of opponents are common currency. In 2001, British polemicist Christopher Hitchens penned *The Trial of Henry Kissinger*, leveling a damning litany of charges against the former US secretary of state, including "war crimes, . . . crimes against humanity, and . . . offenses against common or customary or international law, including conspiracy to commit murder, kidnap, and torture." When other journalists questioned Kissinger about the charges, he dismissed Hitchens as a Holocaust denier. Being of Jewish extraction, Hitchens was understandably enraged by this attack, which Kissinger cynically utilized to turn the focus away from himself.

Yet the *ad hominem* is not always this blatant; often it is well concealed under layers of rhetoric and can require considerable effort to discern. These attacks can be subtle affairs, aiming to cast shadows of doubt over the credibility of the speaker by association with the odious rather than honest engagement with the argument the speaker is making. To see how mischievous and unfair this can be, we need only look at one of the most infamous trials in history: the Papacy versus Galileo Galilei, arguably the father of classical physics. Galileo advanced science by an unprecedented degree, vastly improving the design of the telescope and observing much about the nature of our solar system. His pioneering technology and keen physical insight rapidly led him to the inescapable conclusion that Earth rotated the sun, rather than vice versa.

This wasn't a new idea; it had been proffered by Nicholas Copernicus as a theoretical possibility before his death in 1543, and Galileo rapidly grew convinced of its validity. Yet this was dangerous ground—in the 1600s, the Bible was considered inerrant, incontrovertible by any notions propounded by man. A literal reading of the Bible held that Earth was at the origin of the solar system and all rotated around this perfect point, which was completely at odds with the data Galileo was fast accruing. This was also a time of religious and social strife; the emergence of Protestantism in the late 1500s had fragmented the supremacy of the Catholic Church. The Roman Inquisition had been established to root out any unorthodox beliefs, leveraging the weapon of terror to keep the God-fearing populace compliant. Their conduct was frequently vindictive and bloody, their reach practically unlimited. The penalties for being convicted of heresy included death by burning, and neither men of the church nor men of learning were exempt from the flames. To take one example, the Italian friar, philosopher, and mathematician Giordano Bruno had been burned at the stake in 1600 for heretical views, including beliefs in Copernican theory. This was a travesty of which Galileo was no doubt aware. The mere accusation of heresy was enough to taint a man's character, rendering it unnecessary to deal with anything as inconvenient as his actual arguments. Mindful of the intense religious and political climate, Galileo put forward his views tempered with the gentle suggestion that the Bible's poetry, science, and stories might be interpreted as clever allegories and metaphors rather than as a rigid description of reality.

This careful treading was not enough to shield him from the prying eyes of the faithful. By 1615, Galileo's heliocentric views had been reported to the Inquisition. His interpretation of the Bible as allegorical was seen as borderline heretical. Galileo presented himself to the Inquisition in Rome, but his words fell on deaf ears. In 1616, the inquisitorial commission unambiguously lambasted the heliocentric model to be "foolish and absurd in philosophy, and formally heretical since it explicitly contradicts in many places the sense of Holy Scripture." Under the orders of Pope Paul V, an injunction was served on Galileo to desist from "the opinion that the sun stands still at the center of the world and the earth moves, and henceforth not to hold, teach or defend it in any way whatsoever, either orally or in writing." The Inquisition also banned works by Copernicus, viewing their contents as an insult to faith. While grossly unfair, censure was preferable to some of the other punishments the Inquisition was happy to dole out.

Wisely, Galileo chose not to wade back into the controversy for almost a decade. In 1623, his friend and admirer Cardinal Maffeo Barberini was elected Pope Urban VIII. Barberini had supported Galileo in 1616, and his election seemed promising for academic freedom. With permission from the newly anointed Urban VIII, Galileo was finally allowed publish on the heliocentric system, albeit with some conditions. Chief among these was the caveat that he would not advocate a heliocentric model, instead presenting all views equally. Urban VIII also insisted that some of his arguments for the geocentric model of a stationary Earth be included. The ensuing work, *Dialogue Concerning the Two Chief World Systems*, was a huge success even with these constrictions. It was presented as a conversation among Salviati (a heliocentrist), Sagredo (an initially neutral layman), and Simplicio (a geocentrist). While the book is ostensibly a conversation outlining both views, there is no doubt in the text as to which theory is better.

The name Simplicio for the geocentrist was a thinly disguised insult. Ostensibly named after sixth-century philosopher Simplicius of Cilicia, the connotation of "simpleton" was apparent and deliberate. In an artistic flourish, Galileo modeled aspects of the disagreeable and dull-witted Simplicio after his biggest detractor, conservative philosopher Lodovico delle Colombe, leader of a Florentine contingent of Galileo's opponents, derisively referred to by Galileo and his friends as the "pigeon league." Had this been the extent of Galileo's subversiveness, he might have only ruffled the feathers of these harmless if cantankerous birds. Yet, in keeping his promise to Urban, Galileo inadvertently drew the wrath of the papacy—and the Inquisition. True to his word, Galileo had indeed included the pope's arguments for Earth's primacy in the heavens, sometimes verbatim. The problem is that he put them into the mouth of Simplicio, whose idiocy was clear to any reader.

This soured Galileo's friendship with the pope. Sale of the tome was immediately banned, and Galileo was again hauled before the Inquisition. This time, however, he was accused of heresy, threatened with torture, and held under arrest. An accusation of heresy was not only a grave insult in the 1600s; it was an immensely dangerous stigma to bear. This was the height of the Inquisition, where such pejorative terms could threaten one's social standing, and even one's life, in the grisliest of manners. But more than this, the implication was that heretics were untrustworthy, their ideas devoid of merit. By attacking Galileo as implicitly dishonest and tainted by sin, the papacy never had to counter the actual argument being made. Rather than justify their position or refute anything, the Inquisition simply denigrated the speaker, discrediting him with the mark of heresy.

In 1633, the Inquisition ruled that "the proposition that the sun is the center of the world and does not move from its place is absurd and false philosophically and formally heretical, because it is expressly contrary to Holy Scripture." Galileo threw himself at the mercy of the Inquisition, but it was too late; he would spend the remaining years of his life under house arrest. Pope Urban's dislike of Galileo and the stigma of heresy never faltered. After Galileo's death in 1642, the embittered pope refused permission for him to be buried with his family. Yet, despite all attempts by the Inquisition to smear Galileo, the overwhelming evidence for the heliocentric model eventually vindicated him. His books were finally taken off the Index of Forbidden Books in 1835, more than two centuries later.

A variation of the *ad hominem* fallacy manifests in *tu quoque* ("you, too") exchanges, where the retort to an argument is to accuse the speaker of engaging in the same behavior. This might be evidence to establish the speaker as a hypocrite, but it does not necessarily diminish the validity of their argument. For example, a long-term smoker might implore their offspring to not take up their highly addictive habit, citing the damage it does. The teenager might balk at this apparent double standard, convinced that the parent's habit invalidates their position. But this would be mistaken, as personal inconsistency does not invalidate an argument—in this case, the argument that smoking is damaging still stands, even if the person putting forward this argument sustains a habit that would put a chimney to shame. A related tactic is "poisoning the well," where unfavorable information about an individual (real or fabricated) is used preemptively to discredit a speaker, even if that information has nothing to do with the topic at hand.

Given that arguments from nature are so inherently shoddy, why then do they have such intrinsic appeal? Part of the reason might be that we, like mathematicians, are fundamentally essentialists at heart. Humans are prone to what social psychologists refer to as the *fundamental attribution error*. This is the observation that we place undue emphasis on personal characteristics (intention and character) for the actions of others, rather than contemplate external or situational factors. The person who cut us off in traffic must have done so because they are selfish. We don't tend to think that maybe it was an honest accident or that they were rushing someone to the hospital. Conversely, when we err, we are far more likely to blame circumstances for our actions—we cut that person off because we were late for an appointment.

Similarly, many of us can ignore the plight of homeless and destitute people by convincing ourselves they have some intrinsic flaw, rather than consider the uncomfortable idea that their situation heavily depends on social and economic factors beyond their control. We have an overarching tendency to believe that the bad actions or luck of others are because they are bad people, without considering situational factors that might have played a role. Arguments from nature are too often employed as a comfortable rationalization for indefensible behavior and poor reasoning. Natural fallacies have smeared entire ethnic groups with some ostensible mark of Cain, leading to bloody and oppressive consequences. After all, if you want to justify treating someone as subhuman, the assertion of some intrinsic failing is a powerful way to dehumanize them.

Finding viable solutions to the issues we face together as a species means we must be wary of knee-jerk essentialism. People and situations are intrinsically complicated; simplistic concepts like "good" or "bad" do not map well to either people or ideas. When confronted with ideas or situations, we must strive to avoid guilt by association with preconceived prejudices. We must assess ideas on their own merits and avoid tarring them with the same brush when it's not appropriate. Otherwise we reduce complex issues to farce, and people with all their nuances to two-dimensional heroes or villains. If nothing else, this might be the impetus to be kinder to each other in a world with multitudinous people and divergent views.

7

BAIT AND SWITCH

*The Flawed Rhetoric of Anti-Evolution Theories,
Circular Reasoning, and Cannabis Quacks*

When *On the Origin of Species* was published in November 1859, it was an unexpected hit. The initial run of 1,250 copies sold out within a day of its publication. Charles Darwin's book introduced the world to the theory of evolution by natural selection, a pinnacle of scientific achievement. Crafted for nonspecialist readers, it beautifully laid out the thesis that species evolve over long periods of time in response to selection pressures of their environment. This hypothesis was supported by swaths of evidence accumulated on his expeditions, pointing toward a wonderful and humbling truth: the sheer diversity of life on this planet arose by descent from a common ancestor. Darwin's insight revealed that every species, both extant and extinct, were branches on a diffuse tree of life, intimately connected to all other life on Earth. In the many decades since Darwin's classic was first published, the evidence for evolution has simply become overwhelming.

It would be difficult to overstate the impact of Darwin's work today, as it forms the cornerstone of modern evolutionary biology. At its core, Darwin's idea is elegant and powerful. In a given population, random mutations result in significant variation between individuals. These traits are often heritable, passed down from parents to offspring. When competition for food and resources is intense, individuals less suited to the environment are less likely to be able to survive and reproduce. Conversely, individuals better suited to circumstances have a higher likelihood of reproducing, passing on their traits to future generations. This is the process of natural selection, and slowly, over time, variations and divergence accumulate, leading to the formation of new species. The philosopher and biologist Herbert Spencer called this "survival of the fittest," a term Darwin and fellow pioneer of evolution Alfred Russel Wallace adopted, with the hope of avoiding any misconception that nature itself was actively making selections.

But while Darwin and his contemporaries strove to circumvent misunderstandings, confused interpretations endure even to this day. Evolution remains perennially controversial in certain quarters. By 1860, natural selection was the hottest topic of discussion in Victorian London, and Darwin himself was dogged by the misapprehensions of others about his work. With intense public interest came inevitable backlash. Natural selection rendered man part of the animal kingdom rather than some superior being apart from it. This jarred the religious sensibilities of many senior Anglicans, who saw this and the transmutation of species as an affront that removed God from the process of creation. Even Darwin's former geology mentor, the Reverend Adam Sedgwick, rejected the hypothesis outright, warning that unless his old friend accepted the infallibility of the Bible they would never meet in heaven. Darwin's brilliant insight also incurred the displeasure of an immensely powerful enemy—Richard Owen.

Owen was a towering figure in British science, a skilled anatomist and natural philosopher whose achievements included coining the term dinosaur and driving the creation of the British Natural History Museum. Despite these laudable achievements, he was also capable of being conniving and vindictive. One especially ugly demonstration of this was his pillaging and character assassination of the brilliant paleontologist Gideon Mantell. Perhaps spurred by jealousy, Owen used his lofty position in British science to suppress Mantell's pioneering research papers and had the audacity to present these discoveries as his own. When the perpetually tragic Mantell suffered a carriage accident that left him permanently paralyzed, Owen wasted no time in stealing more of his credits and renaming the specimens Mantell had diligently pored over. Mantell lamented that it was "a pity a man so talented should be so dastardly and envious." Even when the morphine-addicted, destitute Mantell died in 1852, Owen wasn't quite finished denigrating his luckless opponent, penning an anonymous obituary that dismissed him as a mediocre scientist who had done little. Compounding this injustice, Owen had a section of Mantell's spine removed and displayed in the British Museum.

Although Owen's fellow researchers were appalled by his behavior, he remained a powerful individual with noxious reach. Focusing his ire on Darwin, he resorted to his signature smear tactics, penning an anonymous poisonous review of Darwin's work in the *Edinburgh Review*, heaping praise upon himself in the third person. As Darwin's fame grew, Owen grew ever more antagonistic. Darwin himself echoed Mantell's earlier assessment: "Spiteful, extremely malignant, clever; the Londoners say he is mad with envy because my book is so talked about ... it is painful to be hated in the intense degree with which Owen hates me."

Darwin's own failing health kept him from debating the merits of his work against the scores of detractors, but his position was taken up by an emerging band of scientists and philosophers unafraid to oppose clerical thought. One of these was Thomas Henry Huxley, a fine anatomist and visionary in public scientific education. While initially skeptical, Huxley quickly came under the sway of Darwin's elegant idea and the meticulous evidence presented, and he began to argue robustly in support of natural selection. When Owen's fragile ego was dented by Huxley's powerful public ripostes, he defaulted to underhanded subterfuge. Unable to counter Huxley's argument, Owen dismissed him as an "advocate of man's origins from a transmuted ape," a sentiment that rankled the sensibilities of orthodox Victorians against Darwin, turning the sickly scholar into a lightning rod for controversy.

This was mischievous in the extreme on Owen's part, crafted to imply that man had sprung forth from our modern ape cousins. Yet one does not require a doctorate in evolutionary biology to see this is a corruption of what Darwin actually proposed. Natural selection lent weight to the idea that apes and humans shared a common ancestor far back in time, but most certainly did not suggest humans are descended from modern apes. Owen was fully aware that his deliberate implication would trigger predictable emotive reactions, deviously imbuing it with enough cosmetic similarity to the actual substance of Darwin's thesis to fool the unwary. This rendered it an especially potent but intellectually vapid attack. Critics of Darwin mocked him by perpetuating this gross distortion of his work, and he was frequently depicted in mocking cartoons with the body of a monkey.

On June 30, 1860, an infamous debate on Darwin's theory was held at the Oxford Museum of Natural History between a number of prominent supporters and detractors. Leading the opposition was the Bishop of Oxford, Samuel Wilberforce. While a great public speaker, his fawning manner did not endear him to everyone; Prime Minister Benjamin Disraeli derided him as "unctuous, oleaginous, saponaceous," spawning the nickname "Soapy Sam." The night before the debate, Owen schooled Wilberforce in the tactics to employ. And sure enough, Owen's disingenuous ploy arrived in the heat of the debate, when Wilberforce mischievously enquired of Huxley if he was descended from apes on his grandfather's or grandmother's side.

Huxley, who had not earned the nickname "Darwin's Bulldog" for nothing, verbally eviscerated Wilberforce, replying unflustered: "If ... the question is put to me would I rather have a miserable ape for a grandfather or a man highly endowed by nature and possessed of great means & influence & yet who employs these faculties & that influence for the mere purpose of introducing ridicule into a grave scientific discussion I unhesitatingly affirm my preference for the ape." After Wilberforce's loaded jibe and Huxley's acerbic skewering, the debate descended into farce. This culminated in the bizarre spectacle of Darwin's former *Beagle* traveling companion, Admiral Robert FitzRoy, shaking a Bible of immense proportions at the audience, imploring them to accept God over man.*

Owen's devious tactic of distorting Darwin's argument is an archetypal example of a *straw man argument*. This gambit at its most basic is a "bait-and-switch" tactic, pivoting on the impression of refuting an opponent's argument while relying on an easily defeated substitute in its stead. The straw man gambit is particularly well named, conjuring up the evocative image of a sword-fighter demonstrating his prowess on a hay-stuffed mock-up in lieu of an opponent capable of parrying his blows. While there is no challenge in defeating an effigy, this style of argument can seem rather convincing if the switched proposition bears some superficial likeness to the real argument. This isn't always malicious in origin—a straw man attack might arise from ineptitude, if one mistakenly conflates two different ideas. The great mathematician and philosopher Bertrand Russell alluded to this frequent problem, wearily observing that "a stupid man's report of what a clever man says can never be accurate, because he unconsciously translates what he hears into something he can understand."

Straw man gambits are also intentionally employed for more devious purposes, occupying pride of place in many an orator's arsenal. Misrepresenting an argument makes it easier to undermine, and examples of this tactic are legion in every walk of life. Indeed, to find an illustration of this tactic all one has to do is pick up a newspaper or endure a tedious political exchange between two opposing parties, or—heaven forbid—dip a toe into the tumultuous world of online discourse. By their very nature, such arguments are vapid and should succumb quickly to any nonpartisan analysis. Yet frequently the damage is done by conflating a reasoned argument with an emotive misrepresentation, invoking outraged,

---

* Nowadays outside the museum and adjacent to the science library in Oxford, there stands a plaque to that famous debate. The building also houses the anthropological Pitt-Rivers Museum, and combined they form perhaps my favorite exhibition space in the city, as charmingly eccentric as the cast of characters who thundered in its halls, housing everything from dinosaur fossils to shrunken heads.

and disgusted reactions. The unfortunate conflation cemented in the public mind, these visceral reactions can persist and preclude rational discourse.

Sadly, evolution remains a prime target for such empty rhetoric. Since that boisterous debate in 1860, overwhelming evidence has completely vindicated Darwin's idea and evolution is a fundamental principle of biology that has withstood the multitude of challenges to which it has been subjected. But, over 150 years later, straw man objections to evolution continue apace in religious communities that reject natural selection. Whether through ignorance or deliberate malice, variations of the theme "if humans descended from monkeys then why are there still monkeys?" abound, even though evolution does not state that man arose from modern monkeys, and nor is there anything in evolutionary theory that demands a particular source population must go extinct in order for a new species to arise.

On the fringes of devout religiosity unconcerned with objective fact, falsehoods persist with wild abandon. In 2007, evangelists Ray Comfort and Kirk Cameron dumbfounded the scientific community when they insisted evolution must be wrong because no one had found the fossilized skeleton of a hybrid like the croco-duck, holding aloft a poorly photoshopped crocodile-duck hybrid. As attempts to undermine evolutionary theory go, this must rank among the stupidest imaginable, lending weight to Russell's dictum on the fundamental inaccuracy of "a stupid man's report of what a clever man says." The great evolutionary biologist and science writer Richard Dawkins's exasperated reply to such foolishness was understandable and perfectly cutting—in correspondence with me he said: "If your understanding of evolution is so warped that you think we should expect to see a fronkey and a croco-duck, you should also wax sarcastic about the absence of a doggypotamus and an elephanzee ... plus a billion other misshapen chimeras, every living species combined with every other living species."

Sadly, empty arguments of precisely this caliber remain something of a defining feature of the creationist movement—a fact that speaks volumes about the intellectual integrity of anti-evolutionists. In 2001, US State Representative Sharon Broome of Louisiana proposed a resolution that would have condemned "Darwinist ideology" as racist, stating:

> Be it resolved that the legislature of Louisiana does hereby deplore all instances and all ideologies of racism, does hereby reject the core concepts of Darwinist ideology that certain races and classes of humans are inherently superior to others, and does hereby condemn the extent to which these philosophies have been used to justify and approve racist practices.

This of course is an utter butchering of the careful argument that Darwin laid out. Such inflammatory claims are unfortunately common from the anti-evolution movement, deliberately constructed to appeal to emotions and tarnished by flimsy association.* Being on the receiving end of a strawman attack is always frustrating, but it can be more sinister than that. If the disingenuous argument is sufficiently inflammatory, it can be twisted to paint targets on the wrong people or to inure dubious arguments from criticism.

Take, for example, cannabis, used for both recreational and medicinal reasons for millennia. While cannabis has long been with us, few substances excite discussion quite as potently. Explosive claims about its curative power circulate wildly online—a quick Google search for "cannabis cures" yields anecdotes of miraculous efficacy for every illness imaginable. Cannabis heralded as a panacea for cancer is especially common, alongside claims it can alleviate epilepsy or autism. Gushing testimonials aside, this framing of cannabis as a universal panacea is distinctly at odds with the evidence. A 2017 review by the National Academy of Sciences looked at over 10,000 studies on medical application and efficacy of cannabis and cannabis-derived products, finding reliable evidence

---

* Dawkins reminded me that low blows such as this were countered by the urbane wit of Darwin's contemporary Benjamin Disraeli: "Is man an ape or an angel? My Lords, I am on the side of the angels."

for three distinct applications. First, there is strong evidence that tetrahydrocannabinol (THC, the chief psychoactive ingredient in cannabis) can reduce nausea and vomiting associated with cancer treatments. These anti-emetic properties have been exploited for decades in the clinical management of cancer symptoms. The review also found good evidence to support the use of medicinal cannabis in chronic pain, and in managing spasms associated with multiple sclerosis. There were, however, caveats to these findings. THC itself is not uniformly well tolerated, leading to many instances where it exacerbates rather than placates vomiting. As safer and more efficient medications and painkilling agents exist, clinical compounds derived from THC tend to be reserved only for when other interventions have failed.

But what of the breathless claims of universal curative potency preached by cannabis advocates worldwide? On this, the same review was more damning. Despite the unending hyperbole, the evidence for the efficacy of cannabis in other conditions was minimal and unconvincing. The authors could find no convincing evidence that cannabis was useful for the treatment of ADHD, epilepsy, Parkinson's disease, irritable bowel syndrome, or appetite management for AIDS patients. Most certainly there was absolutely no evidence that cannabis can treat or cure cancer. Sean Hennessy, one of the study's authors, reflected on this gulf between evidence and belief, observing that "most of the therapeutic reasons people use medical marijuana aren't substantiated beneficial effects of the plant." For unrelenting supporters of medicinal cannabis, though, the relatively modest and limited application of the drug has been no impediment to their advocacy, nor has it moderated their contentions.

The "cannabis cures cancer" myth staggers on, zombie-like, with a depressingly huge number of websites evangelizing that fiction. Across social media sites, a worrying number of posts circulate endlessly, claiming cannabis oil or THC extracts cure cancer. Often these are pitched at patient support groups, at vulnerable cancer patients and their families. These claims are ostensibly maintained

by an entire family of straw men, exemplified by a common assertion
made to "prove" that cannabis cures cancer: High-dose THC can kill
cancer cells in a petri dish. This is true—but irrelevant. Killing cells
is simple; plated cancer cells are easy to eliminate by many agents,
from acid to heat to bleach. The astute reader will note, however,
that humans are not petri dishes. Effective anti-cancer agents for
us must be able to target cancer cells discriminately while sparing
healthy ones. There is no evidence whatsoever that cannabis can do
anything of the sort. So enduring are these falsehoods that organizations like Cancer Research UK and the National Cancer Institute
have had to dedicate ample resources to tackling these myths.

At best, such misrepresentations offer a superficial veneer of
credibility for views incongruent with the evidence. But at worst,
they can do serious damage to at-risk patients. Despite the ubiquity
of cancer, the general public is still widely underinformed on the
topic, preferring to avoid the issue or referring to it with deferential
cloaked euphemisms. Most people don't give cancer serious thought
until they or a loved one are diagnosed. While modern cancer treatments and survivability are constantly improving, the prospect of
interventions like radiotherapy, chemotherapy, or immunotherapy
is understandably frightening. When patients are at their most
vulnerable, the allure of "natural" elixirs with no side effects can be
so tempting that it overrides healthy skepticism. More insidiously
perhaps, such claims foster mistrust between patients and the medical and scientific community. To preserve their faith, true believers
dismiss the lack of evidence for their position as the machinations
of "Big Pharma." It might be tempting to dismiss this as a crackpot
view but, as we shall see later, this is a widely held belief.

The inevitable result of accepting these fictions is to cast researchers and medics as villains, marking them as targets for abuse and
scorn. Worse than that, the miracle-cure narrative is delivered so
fervently that it persuades many patients to cease their conventional
therapies. This has—and will continue to—cost lives. I can personally
attest to these aspects. In 2016, a bill came before the Irish parliament,

ostensibly advocating for medicinal cannabis, despite medicinal cannabis already being available in Ireland subject to prescription. Reading
beyond the stated purpose, the true purpose of the proposed bill
became clearer: unfettered access to the drug for conditions where
it had no demonstrable efficacy. More than that, it touted cannabis as
a medicine while simultaneously excluding it from the remit of the
national drug-regulation body. Essentially this was Schrödinger's bill,
asserting that cannabis was medicinal but insisting it not be subject
to the regulation or control this would normally entail.

This nonchalance was a serious red flag that raised the suspicions
of the more scientifically inclined. Anything with biological or
medicinal impact is likely to have some potential ill effect, and cannabis is no exception. Though cannabis is relatively safe, current data
suggests that regular users have an elevated chance of mental-health
disorders, including schizophrenia. These effects appear much more
pronounced in children and adolescents, with negative implications
for educational and social attainment. Contrary to popular misconception, cannabis addiction is entirely possible and problem usage
is common, more likely to manifest in heavy users and those who
begin at a younger age. In this light, the bill's insistence on bypassing
medical regulation was perhaps telling of a different motivation. For
all its ostensible medical focus, the bill was a Trojan horse for the
decriminalization of cannabis for recreational use. That it referred
to patients as consumers, and measured dosage in ounces rather
than the more conventional medical measurement of milligrams,
only amplified this perception.

This isn't necessarily a bad thing—there *are* some excellent
arguments for the legalization of cannabis. Yet it is completely unacceptable and utterly disingenuous to make dangerously misleading
medical claims to this end. By framing it as a medical issue, the
promoters of the bill aligned themselves with the army of cannabis
cranks worldwide. To salt the wound, the tone of the campaigning
and publicity was irresponsible to an almost impressive degree. For
a public meeting in support of the act, People Before Profit (PBP,

the bill's proposers) unveiled a poster stating that cannabis cures cancer. Across social media, people were invited to post their cannabis cure success stories, and sure enough miraculous anecdotes aplenty ensued. That none of these were in any way substantiated (and were frequently debunked) was no impediment to their warm reception. One popular meme in 2017–18 featured the wondrous tale of David Hibbitt, alleged to have cured his cancer with cannabis oil. While cannabis enthusiasts widely shared the meme, they were unaware or unconcerned that—far being cured—Hibbitt had succumbed to his cancer in 2016.

Heart-wrenching emotional appeals were a staple of the campaign at rallies and in the press. The plight of one mother to get cannabis oil for her sick daughter garnered ample media coverage. This was heavily leveraged as an example of why the new bill needed to pass, despite there being no evidence that cannabis oil had any efficacy for the condition in question. In the huge outpouring of sympathy and outrage, the fact that cannabis oil was considered a foodstuff, legal and easily available in Ireland, seemed to escape the public mind. With increased publicity, the "cancer cures" narrative clawed its way deeper into public discourse. Robert O'Connor, head of research at the Irish Cancer Society (ICS), wearily observed: "Everywhere I go now I'm asked about cannabis or CBD oil curing cancer. These false claims have become so common in media, especially social media, as to be dogma in the mind of many in the public, even though the research clearly and unequivocally shows that it's just not true."

Predictably and depressingly, vulnerable patients were targeted by the well-meaning but misguided, showered with glowing anecdotes about tumors shrinking away after cannabis treatment. Yet, as the ICS and oncology professionals kept stressing, these claims were complete falsehoods. As public misconceptions grew, I spoke to numerous media outlets about why these claims were suspect. I authored pieces for *The Irish Times* and *The Spectator*, debunking several claims made both in the bill and by its supporters online, imploring readers to be guided by evidence and not assertion. The Irish parliament convened a bipartisan committee to assess the bill, which returned its findings in July 2017. Its conclusions were damning, exposing a plethora of major legal issues, unintended policy consequences, and a lack of safeguards against harmful use. The medical motivation proffered was specious at best too. The committee unanimously rejected it, deeming it to be "as much about decriminalizing the use of cannabis as it is about promoting it for medicinal use."

This stark rejection left the bill up in smoke, unable to proceed—a failure born of a heady cocktail of sheer ineptitude and outright duplicity. Rather than reflecting on the committee's comments, PBP's Gino Kenny thundered that the bill was "sabotaged" by the committee, lambasting them as biased. Across social media, the response by many of the bill's supporters was to go on the attack against the most prominent critics. Committee member Kate O'Connell was singled out for a flurry of abuse, her qualification as a pharmacist cited by her detractors as "proof" she was in league with the nebulous Big Pharma. As is sadly often the case with women online, many of the comments leveled toward O'Connell were misogynistic in the extreme.

Robert O'Connor and the ICS were accused by the same zealots of being pawns of the drug industry, leading to the unedifying sight of a cancer charity being attacked by cannabis trolls. I didn't escape unscathed either—my ostensible motivations were furiously pored over, abusive messages arriving in torrents. As a testament to the apparently amorphous nature of "Big Pharma," I too was labeled as being under their nefarious auspices, despite being a physicist by qualification. More hurtful perhaps was the constant accusation in the voluminous screeds that our criticism demonstrated lack of compassion for the sick and contempt for the suffering. This was more straw-manning, labeling bill critics as saboteurs eager to undermine a wonderful health revolution—the insistence being that those who didn't support the bill didn't care about patients. This was a complete misrepresentation of what had been said and of the careful consideration that led to the bill's rejection by the committee. In reality, the proposed legislation would have not safeguarded patients but put them in jeopardy. To compound that, PBP had been supremely reckless in pushing the narrative of cannabis as a panacea. By misrepresenting the objections against the bill, the proposers had deflected attention from the fact that the bill itself was dangerous, flawed, and disingenuous. Instead they had painted a target on those with valid concerns, garnering them the misplaced ire of supporters.

The one silver lining to this debacle was that it led to a serious conversation about the ubiquity of outlandish therapies pitched at vulnerable patients. To try to stem this to some extent, Kate and I and others began working on a bill to protect cancer patients from suspect cures. Inevitably perhaps, once we announced this project, it was itself straw-manned by cannabis advocates and purveyors of alternative medicine as "suppressing cures for cancer," leading to all of us getting copious volumes of abuse, with O'Connell observing that: "Obviously in politics you get abuse all the time. But the level of vitriol over this . . . is stratospheric."* This at its heart is the real devious power of the straw man argument. It not only inures vapid positions from valid criticism, it facilitates the outright demonization of those who might point out that the emperor is indeed naked. In this respect, it deals double the damage, and it's crucial we keep on our guard against it.

Most of the informal fallacies we have encountered thus far are variations on the theme of *causal fallacies*. These hold pride of place in the pantheon of the informal, existing in seemingly infinite form. It's worth looking at a class of informal fallacies that rely on flexible definitions and premises to achieve a veneer of rigor. The simplest way to achieve this end is to put forward an argument that

---

* When this was originally written, the "Treatment of Cancer (Advertisements) Bill 2018" was progressing through Dáil Éireann, the Irish parliament. Following a fresh election in 2020, its progress was halted. How and when this is revived remains an open question, but it illustrates the reality that fictions travel faster than legislation.

is completely logically airtight but wholly tautological. Consider the statement: "Humans are mammals; therefore, humans are mammals." This is quite transparently a poor argument, as the speaker is essentially beginning with their conclusion to justify their conclusion. Such fatuous reasoning is known as *circular reasoning*.

While this might seem obnoxiously obvious, it has impressive historical pedigree. To cite but one example, arguments concerning the veracity of religious texts have drawn strongly on the pleasing but vapid circularity of such logic. This has long afflicted the Abrahamic religions—since the dawn of Judaism it has been argued by religious scholars that the Torah is the unerring word of God because the scripture in the Torah says this. Such empty reasoning was adopted enthusiastically by the religions that Judaism gave rise to. To take but one biblical example, we need only look to Timothy 3:16, which states: "All Scripture is given by inspiration of God, and is profitable for doctrine, for reproof, for correction, for instruction in righteousness." This passage, stripped of its verbosity, could be readily rephrased without loss of generality as: The scripture is true because it is inspired by God because it is written in scripture. This might be a reassuring sentiment for dedicated believers, but it rings somewhat hollow to the more astute. Similarly, the Qur'an asserts that the revelation of the holy text to the prophet Muhammad vindicates its divine origin. Whether God exists or not, these patently circular arguments do nothing to advance that hypothesis.

Theological examples are relatively transparent, but circular reasoning can often be obscured by synonyms and complex phrasing that requires some unpacking. This is frequently seen in a related fallacy known as *begging the question*.* In question-begging, the conclusion whose proof is being attempted is contained in the premises of the argument, rendering the entire statement an exercise in tautology. On deep-seated ideological issues, it can be readily abused,

---

* Pedant's warning: "Begging the question" is the logical fallacy outlined here. The phrase is frequently used incorrectly to mean "raising the question." Context tends to clarify what is meant, but it's a pet peeve of mine.

as the begged question often aligns with the views being courted. To take a contemporary example, we might look at the rhetoric of the ever-contentious abortion debate. A common argument against abortion lays out its logic as: "Abortion is murder. Murder is illegal. Thus, abortion should be illegal." This might seem like a convincing argument to those toward the anti-choice end of the spectrum, but as arguments go it is hopelessly flawed. The conclusion that abortion should be illegal stems from the assertion that abortion is murder, an incredibly dubious claim. The conclusion of the argument is entailed in its premises, with the premise begged that we grant the assumption that abortion is murder without any reasoning or evidence as to why that is the case. If the begged premise is allowed, then the entire statement is logically consistent but a pointless exercise in profoundly circular reasoning that tells us precisely nothing.

This rather effectively brings us to an important question, and something of an elephant in the room. We've touched upon some of the more common informal flaws in reasoning, but it's worth stopping for moment to consider a common theme running through many of the examples we've discussed: the struggle between logic and belief. The influence of belief on reasoning is a pertinent one that we have until now skirted. Undoubtedly, faulty reasoning can affect even the most neutral unwary observer, but the motivation is often rather more suspect. In some instances, logically fallacious reasoning might result from simple misunderstanding, but we should not discount the possibility that our beliefs may pose a noxious influence on our ability to think straight. Do we then filter arguments through a skewed lens, giving undue credence to those who support our preconceived notions? And are we more likely to engage in fallacious reasoning to buttress an article of faith we hold dear? And if we do, is this a deliberate act or something of which we're consciously unaware? If we're truly to understand the challenges to reason we cannot look at logic divorced from the intricacies of the human condition. To truly comprehend why we err, then, it's vital that we explore the human traits that exert an inescapable influence on all of us.

PART 3

Trapdoors of the Mind
The Struggle Between Reason and Belief

"Our own faults are those we are the first to
detect, and the last to forgive, in others."

—LETITIA ELIZABETH LANDON

8

SCHRÖDINGER'S BIN LADEN

Stalin's Russia, Climate Change Denial, UFO Cults,
and the Irrational World of Motivated Reasoning

The early twentieth century in Russia was a tumultuous time. The October Revolution of 1917 saw revolutionary Bolshevik forces establish the world's first communist nation. This huge transition created political vacuums, eagerly filled by power-hungry and often unscrupulous men. Joseph Stalin is doubtlessly the most infamous occupant of this rogues' gallery. His vaulting ambition was apparent to the ailing Vladimir Lenin, hero of the revolution and head of the Soviet government. The wary Lenin disavowed Stalin, recommending Leon Trotsky as his heir. Even so, following Lenin's death in 1924, Stalin outmaneuvered all rivals, consolidating absolute power. Trotsky was forced into exile and eventually executed in Mexico on Stalin's orders, bludgeoned with an ice axe. The brutality of Stalinism is well covered in history books, but more obscure is the story of another ambitious man of the era—Trofim Lysenko.

Lysenko's passion was plants rather than politics. While his peers were shaping the revolutionary effort, he was studying seeds in Kyiv under his mentor, Nikolai Vavilov. Their primary interest was investigating the conditions that influence wheat crop yields. This issue rapidly acquired political urgency once the new Russian leaders began to force a rapid transition from an agrarian economy to an industrial one. Affluent "kulak" peasants were eradicated as "class enemies," their fertile land taken over by collective farms. Chronic mismanagement ensued and famines erupted all over Russia due to Soviet ineptitude.

Lysenko's announcement in 1928 that he had found a new way to hugely increase crop yield, which he called "vernalization," was music to the ears of the Communist Party, and the Soviet mouthpiece *Pravda* lavished praise on him. The party's propaganda relied heavily on inspiring stories of ingenious workers solving practical problems by wits alone, so this agronomist from peasant origins without any formal scientific training outsmarting the bourgeois scientific establishment was a godsend. He was bestowed with awards, both political and scientific, and elevated up the hierarchy of the Communist Party. Such praise was premature; Lysenko's lack of scientific training translated into poorly controlled, subpar experiments. Nor was he above supporting his claims with fraudulent data to bolster his heroic image.

Still, suspect results seemed no impediment and accolades continued unabated. Lysenko was an unimpeachable darling of the party. He had come to believe his own hype too, insisting that the offspring of seeds treated with his vernalization process would themselves inherit wondrous properties, so that rye could transmute into wheat, and wheat to barley. This caused consternation to biologists, as it pivoted on something long experimentally refuted: Lamarckian evolution. This obsolete theory suggested that acquired characteristics of an organism could be passed down to its descendants. For example, a plant that had been plucked of leaves might have a leafless offspring. Biologist Julian Huxley pithily observed that "if this theory is correct it would follow that all Jewish boys would be born without foreskins."

Unlike Lysenko, most Russian botanists and biologists had been educated prior to the revolution. They were familiar with Darwin's theory of evolution, which provided a richer explanation of what was observed and had passed the trial by fire of intense experimentation. They were also aware of the ideas of Gregor Mendel and experiments with fruit flies that suggested a unit of heredity—the gene. Yet in political ascendency, Lysenko did not acquiesce one iota to the concerns of scientists. Unable to refute their observations, he resorted to ad hominem attacks. In 1935, he compared those who did not accept his ideas as indistinguishable from those who rejected Marxism, denouncing biologists as "fly-lovers and people-haters." After this outburst, Stalin himself was the first to stand in ovation, bellowing "Bravo, Comrade Lysenko. Bravo."

Praise emboldened Lysenko further. With explicit approval from Stalin himself, he began melding his agricultural ideas with those of communism. The entire field of genetics became his primary target; an interpretation of Marxist doctrine suggested that human characteristics could themselves be changed by living under communism. These acquired improvements would be inherited by subsequent generations, creating a heroic "new Soviet man." This was a far more politically tractable belief than the alternative—that one's traits are largely shaped by unalterable genetic code, so that rye can never become wheat. Lysenko thus rejected Darwin's work on competition in natural selection, decrying the very idea as anti-communist.

As the Second World War consumed Europe, Lysenko, with Stalin's blessing, began to purge those scientists who had contradicted his grandiose claims. His mentor and early champion Vavilov was arrested on overblown charges and sentenced to death, though this was commuted to imprisonment. Vavilov died in prison from malnutrition in 1943. In 1941, Nazi Germany attacked Russia, and a long bloody war ensued that meant Lysenko's crusade was temporarily put on ice. At the war's end in 1945, the Soviet Union emerged victorious, but at a colossal price, with almost 27 million Russians perishing in the conflict. While Lysenko still held massive sway with the party, some scientists had become bold enough to openly criticize his dictatorial influence. A series of evaluations of his work by others showed his claims to be either unjustified or blatantly falsified. Apprehensive of his position, Lysenko implored Stalin for extra support, promising to increase the country's wheat yield up to tenfold. Despite ample evidence that this was impossible and Lysenko incompetent, Stalin bowed to this much-vaunted genius of the proletariat, bestowing the entire political machinery of the Soviet Union on Lysenko.

So in 1948, Lysenko officially declared genetics a "fascist" and "bourgeois perversion." The Politburo declared the entire discipline of genetics prohibited across the USSR; Lysenkoism was now the only "politically correct theory." The party decree attesting to this was edited by Lysenko and Stalin himself. All genetic research was forbidden and further discussion outlawed. Biologists and geneticists across the USSR were unconditionally fired, their work publicly condemned. Approximately 3,000 scientists were rounded up and executed or sent to gulags and prisons. These persecuted scientists in genetics, biology, and medicine were replaced by incompetent syco-phants loyal to Lysenko. To compound matters, Lysenko's backward agricultural policy plunged the country into further starvation.*

This iron hold had a chilling effect on scientific discourse. Stalin died in 1953, succeeded by Nikita Khrushchev, himself sympathetic to Lysenko. Following Khrushchev's dismissal in 1964, the scientific establishment in Russia could finally mount an offensive. At the General Assembly of the Russian Academy of Sciences, nuclear physicist† Andrei Sakharov blamed Lysenko for "the shameful backwardness of Soviet biology and of genetics in particular, for the dissemination of pseudoscientific views, for adventurism, for the degradation of learning, and for the defamation, firing, arrest, even death, of many genuine scientists." In concert with this damning sentiment, reports emerged showing that Lysenko and his acolytes had falsified and misconstrued evidence.

With no political sponsors to protect him any more, Lysenko's house of cards fell to pieces under the spotlight of unrestrained scientific inspection. The death knell had been sounded for his stranglehold on Soviet science. The state press, which had once heralded his genius, now damned him absolutely. Lysenko retreated into obscurity, dying quietly in 1974. His cult of personality had stifled advances in genetics, biology, and medicine across the Soviet Union, his peaceful end standing in stark contrast to that of the

---

* Tragically, many of Lysenko's beliefs were adopted in China, exacerbating the Great Chinese Famine encountered in the introduction.

† Curiously, the only scientists capable of criticizing Lysenkoism without grievous consequence were physicists. Historian Tony Judt remarked: "It is significant that Stalin left his nuclear physicists alone and never presumed to second guess their calculations. Stalin may well have been mad but he was not stupid."

scientists whose destruction he had authored in his violent purges. The Lysenko affair was, in the words of scientist Geoffrey Beale, "the most extraordinary, tragic and in some ways absurd, scientific battle that there has ever been."

But Lysenko's story is more than just the hubris of one man; it also tells us something about the human condition. The very reason his work was so revered was because it jibed with an ideological stance. It bears the hallmarks of an all-too-human psychological error known as *motivated reasoning*, where evidence—instead of being evaluated critically—is deliberately interpreted in such a way as to reaffirm a preexisting belief. It is an emotionally driven, and inherently biased form of decision making. It demands impossibly stringent standards for any evidence contrary to one's beliefs, while accepting uncritically even the flimsiest evidence for any ideas that suit one's needs. Rather than rationally evaluate evidence that might confirm or deny a belief, motivated reasoning uses our biases to look only at evidence that fits what we already believe and to dismiss that which unsettles us.

Motivated reasoning is closely related to *confirmation bias*, our tendency to seek, remember, and frame information in a way that agrees with our preconceived beliefs and worldviews, while minimizing contradictory information. The idea that we have an internal gatekeeper with a propensity to filter information is not a new one; four centuries before the Common Era, the Greek historian Thucydides noted: "It is a habit of mankind to entrust to careless hope what they long for, and to use sovereign reason to thrust aside what they do not fancy." This observation has been supported by a wealth of data by psychologists in the twentieth century who began formally examining the scope of our ability to placate ourselves with convenient fictions. Yet there is a high cost associated with clinging to falsehoods, however comforting—so why would we do this?

This question captured the attention of pioneering psychologist Leon Festinger, who postulated that simultaneously holding two or more contradictory beliefs on a topic might lead to a form of mental agitation. He termed this "cognitive dissonance," the discomfort a person feels when they encounter information or actions that conflict with those they already hold. When confronted with clashing information, we endeavor to quell this discomfort. We might accept that our preconceived notions may be flawed or incomplete, and—like an ideal scientist—refine our views in light of new evidence. But to alter our ideological leanings is cognitively expensive; an easier option is simply to deny reality in order to preserve our beliefs. In Festinger's paradigm, motivated reasoning is a mechanism to stave off discomfort from conflicting information, "motivating" us to accept soothing falsehood over challenging realities.

Festinger derived this notion in the early 1950s and sought a means to test his hypothesis. An intriguing headline in his local paper caught his attention: PROPHECY FROM PLANET CLARION CALL TO CITY: FLEE THAT FLOOD. The article concerned a cult led by Chicago housewife Dorothy Martin, who was adamant that she received communications from the planet Clarion through automatic writing. These alien missives had revealed that the world would end on December 21, 1954. Martin had previously been involved with L. Ron Hubbard's dianetics movement (which later became Scientology) and had cannibalized his B-movie science fiction aesthetic. Martin declared that on the eve of destruction, flying saucers would appear to the faithful to spirit them to Clarion while a great flood hit Earth. Her movement sought spiritual clarity and salvation and adopted a name reflective of this longing: the Seekers.

The specific nature of the Seekers' beliefs made them stand out from the chorus of apocalyptical groups around the US. Martin presented her beliefs in a take-it-or-leave-it fashion; she was not interested in proselytizing and was averse to engaging with the media. Even so, she was surrounded by a small band of devotees who believed so strongly in her proclamations that they had surrendered not only their positions and material possessions but in some cases their marriages and families in order to follow her. Recognizing a unique opportunity to study how strong beliefs fare when faced with incontrovertible evidence that undermines them, Festinger and his colleagues arranged for students to join the group under cover and observe the Seekers first-hand, tasked with recording how they would deal with the inevitable disconfirmation of their beliefs.

Throughout December 1954, the Seekers began preparing for the imminent destruction of the world. They nervously awaited the next communication from Clarion, which came through Martin at 10 AM on December 20; it assured them they would be saved from the destruction and whisked into space. All metal was removed from their persons, lest it wreak havoc on the flying saucers, so bra wiring, zippers, and metal adornments were dutifully removed. More communications came throughout the day, including a series of passwords the saved would need to board the rescue vessel, which was to arrive at midnight. The group drilled these call-and-response codes, convinced they would survive Earth's demise. At 11:15 PM Martin ordered her followers to don their coats, and as midnight approached, the faithful huddled together in silence, awaiting salvation.

But as midnight arrived, nothing was seen. A clock in the room read 12:05, while another read 11:55. With mounting anxiety, the group agreed that the earlier clock must be correct. They waited until it struck midnight, sick with anticipation. But as the hands of the clock aligned, no savior emerged. Over the next hours, a mournful agitation engulfed the room—cataclysm was due to devour Earth by 7 AM, and the promised rescue had not materialized. By 3 AM, the group was desperately picking apart the words of the prophecy, looking for hidden symbolism they had perhaps overlooked. But their attempts to rationalize rang hollow even to themselves. By 4 AM, some were in tears while others sat listlessly in shock. But this dejected mood didn't linger long, for at 4:45 AM

Martin called everyone together to deliver a message just received from Clarion. It stated:

> For this day it is established that there is but one God of earth, and He is in thy midst, and from his hand thou hast written these words. And mighty is the word of God—and by his word have ye been saved—for from the mouth of death have ye been delivered and at no time has there been such a force loosed upon the earth.

The Seekers were ecstatic, convinced they had saved Earth from doom. They had rationalized away the abject failure of their prophecy, instead painting it as a glorious thing. Completely reversing their prior position, they became vocal evangelists and urgently sought media attention.

It's worth noting that the Seekers were not the first, nor the last, group to double-down on their positions despite their great prophecy not coming to pass. The Millerite movement believed Jesus Christ would reemerge in the year 1844, and his failure to appear was dubbed the "great disappointment." Yet this too was rationalized away. As of 2010 the Adventist churches that arose from Millerite beliefs have roughly 22 million followers worldwide. That people can become more fervent believers after tenets of their faith are explicitly refuted might seem strange to us, but it was precisely what Festinger and his colleagues predicted. In their seminal work on belief, *When Prophecy Fails*, they laid down five conditions that must be present for this to transpire:

1. A belief must be held with deep conviction and it must have some relevance to action, that is, to what the believer does or how he or she behaves.
2. The person holding the belief must have committed himself to it; that is, for the sake of his belief, he must have taken some important action that is difficult to undo. In general, the more important such actions are,

and the more difficult they are to undo, the greater is the individual's commitment to the belief.

3. The belief must be sufficiently specific and sufficiently concerned with the real world so that events may unequivocally refute the belief.

4. Such undeniable disconfirmatory evidence must occur and must be recognized by the individual holding the belief.

5. The individual believer must have social support. ... [If] a group of convinced persons ... support one another, the belief may be maintained and the believers may attempt to proselytize or persuade nonmembers that the belief is correct.

Festinger summed all this up later with the pithy observation that "a man with a conviction is a hard man to change. Tell him you disagree and he turns away. Show him facts or figures and he questions your sources. Appeal to logic, and he fails to see your point." This is not solely a failing that afflicts the religious; the needless controversy over climate change is underpinned by a similar rationale. The perception that climate change is scientifically contentious is widespread but mistaken—the scientific consensus is simply overwhelming that we as a species are drastically altering our climate. Part of the reason for this gulf between perception and reality is due to years of skewed coverage, a topic we'll cover in future chapters. The mechanism of action has been well known for over a century; French polymath Joseph Fourier hypothesized a human effect on climate in 1827, and the effect of CO_2 and other greenhouse gases were demonstrated experimentally by Irish physicist John Tyndall by 1864.

That humans can thus affect climate is no surprise; what is surprising is just how fast we're doing it. Ancient ice cores contain a record of temperature and atmosphere spanning hundreds of millennia, showing our current rate of warming to be incontrovertibly hundreds of times beyond anything that has gone before. More alarming is that, while at no point during any previous glacial or interglacial period has the CO_2 concentration level reached as high as 300 ppm (parts per million), in September 2016, we surpassed the 400 ppm threshold, with predictions of up to 600 ppm in the coming decades. This is most distinctly not natural variation. Nor can we evade responsibility by postulating that this level is unrelated to human activities—CO_2 released from fossil fuels bears distinct chemical signatures that point to our guilt as readily as fingerprints on the trigger of a smoking gun. The inescapable conclusion is that we ourselves are driving the destruction of our own environment.

The evidence for this is virtually incontrovertible, and yet there is a sizeable contingent who insist this isn't the case. Self-proclaimed "climate skeptics" scorn the overwhelming evidence that climate change is happening or insist we are not responsible. But these nay-saying "climate skeptics" are engaging in a calculated misnomer. Scientific skepticism is crucial to probing whether a particular hypothesis is supported by evidence, and is a vital element of the scientific process. Yet climate skeptics persist in ignoring empirical evidence that renders their position untenable. This isn't skepticism; it is unadulterated denialism, the very antithesis of critical thought. Accordingly, I refer to such individuals as climate change denialists, not skeptics, in line with scholarly convention. The National Center for Science Education notes too that denialism also encompasses unwarranted doubt as well as outright rejection.

Being bereft of scientific support seems no impediment to this veritable brigade of underinformed armchair experts, however. They lurk on comment threads the world over, downplaying climate science. Climatologists are all too frequently the targets of their ire, as are the journalists who communicate these research findings. Were such passionate disdain confined to the foaming underbelly of internet comments and forums it would be bad enough, yet this attitude is frequently encountered in the press. The editorial position of many tabloid newspapers is often denialist, the Murdoch press being particularly vehement in denying reality.

The schism can be especially apparent in politics, with a depressingly large number of denialists holding office across the world. In 2009, Australian Prime Minister Tony Abbott declared climate change "absolute crap." In the UK, swaths of the UK Independence Party and even the Conservative Party downplay climate change. But political denialism is undoubtedly strongest in the US, where a 2016 survey suggested that a third of Congress members are denialists. The Republican Party is unique among major conservative parties worldwide in that they are the ones to be explicitly denialist. Members such as Senator Jim Inhofe* even argued that climate change was a conspiracy by scientists to garner funding. Perhaps most egregiously, Republican President Donald Trump insisted that climate change was a conspiracy by the Chinese to hobble American industry.

Given that the weight of scientific evidence is so strong against these claims, why then do these beliefs muster such vocal and unrelenting support? Naively, we might assume the problem is simple misunderstanding. Certainly, increasing average global temperature can have paradoxical and counterintuitive repercussions, such as causing extreme cold snaps. Were this the problem, the obvious response would be to articulate the scientific details more clearly and more often. Yet the problem is that this well-meaning and considered "information-deficit" approach hinges on the presupposition that the intended audience is basing their position on the balance of evidence. But as we've seen, if the motivations underlying such vehement protestations are ideological in nature, then such a well-meaning endeavor will always be in vain.

The evidence is ample that rejection of science is often not motivated by reason but by ideology. Climate denialism is far more common among politically conservative individuals with traditional values. The role of ideology in acceptance or denial of climate science has been a persistent research interest of Stephan

---

* Senator Inhofe is famous for a 2015 outburst where he brought a snowball into Congress, insisting it proved that global warming was a scam. Even by the heady standards of climate change denialists, this was a particularly stupid argument.

Lewandowsky and his colleagues. In their fantastically titled study, "NASA Faked the Moon Landing–Therefore, (Climate) Science Is a Hoax: An Anatomy of the Motivated Rejection of Science," they found that subjects subscribing to conspiracist thought tended to reject all scientific propositions they encountered, while those with strong traits of conservatism or pronounced free-market worldviews only tended to reject scientific findings with regulatory implications–namely, climate science.

This pattern has been repeatedly demonstrated, identifying political views as the biggest predictor of climate change denial. It should come as no surprise that the voters and politicians opposed to climate change tend to be of a conservative bent, with a keen belief in free-market ideology. There is further evidence suggesting that the stronger one's belief in the free market, the more likely one is to dispute climate change.* Take, for example those with an active distrust of market regulation; the existence of climate change presents something of a challenge to their ideology. Provided they are not nihilists, then accepting that human activity has consequences for others might be cognitively difficult, forcing them to refine the nuances of their personal philosophy. For many, it is simply easier to quench intellectual discomfort by retreating into naked denial, ignoring or attacking evidence that conflicts with deeply held beliefs.

For those with pronounced free-market views, climate change confounds deeply held belief. For if one accepts human-mediated climate change, then supporting mitigating action should follow. But the demon of regulation is a bridge too far for many libertarians. Given that climate change affects everyone, whether they consent to it or not, then unregulated use of natural resources infringes the property rights of others, rendering it ideologically equivalent to trespass. Thus, the property rights house of cards

---

* All of this is also exacerbated by Machiavellian efforts by lobby groups to confuse the issue, which we'll tackle in chapter 15.

comes crashing down. When faced with this ideological dilemma, some free-market advocates resolve the inevitable cognitive dissonance by simply denying the reality of climate change rather than acknowledging that the axiom they cling to may require revision.

It's worth noting that none of this is intended to dismiss legitimate concerns and questions. There are huge questions over how we best address climate change, and frank conversations on the subject are sorely needed. Ideological blindness on climate change isn't solely the preserve of free-marketers, and we'll see later how those on the opposite side of the political spectrum can harbor equally fallacious beliefs to preserve an ideology. But constructive solutions can only be found when we acknowledge reality; problems cannot be rectified if they are not recognized. In this respect, denialists fall at the first hurdle, dismissing the problem and stifling vital dialogue. The sheer intensity of their protestations is telling—like Festinger's UFO cult, they are unwilling or unable to let their position evolve with evidence. Their fury betrays a stance that is emotional rather than rational. Unable to justify their contention, they resort to shouting down those facts that clash with their perceptions and make them uncomfortable, attempting to drown out the intrusions of reality on their perfect ideology. This would be merely sad, if their sustained assault on reason didn't carry with it serious implications for the future of our very planet.

It is difficult to overstate the influence of ideological lenses on how we perceive the world, but there have been attempts to quantify this. An infamous 2013 experiment at Yale by Dan Kahan and colleagues gave subjects a neutral problem about whether a certain skin cream could alleviate a rash. Subjects were given the data below and asked whether they thought the ointment was effective or not.

	Rash improved	Rash disimproved
Patients who used new skin cream	223	75
Patients who did not use cream	107	21

The question requires a little savvy to answer correctly. Intuitively, people tend simply to pick the largest number, leading many unwary subjects to assume the cream here was beneficial. A more nuanced analysis requires noting that there is more than twice the number of patients in the cream group than in the no-cream group. Accordingly, to answer this question correctly, we'd need to look at the ratios. There were 298 people (223 + 75) who used the cream versus 128 (107 + 21) who did not. In the cream group, 223/298 (or roughly 75 percent) improved while 25 percent got worse (75/298). In the control group, 107/128 (roughly 84 percent) saw their rash improve while 16 percent (21/128) saw a degradation in their condition. When analyzed this way, the conclusion is at odds with initial assumptions—the cream did not improve patient rashes.

Unbeknownst to the subjects, researchers had a more pressing question than lotions in mind. The subjects had been covertly stratified by political alignment into conservative and liberal cohorts. The neutral question on lotions proved to be rather difficult for many regardless of political affiliation, with 59 percent of subjects tested getting the wrong answer. With the mathematically capable subjects identified, the researchers gave another similar problem. But this time, the question concerned a subject that teeters at the very brink of the American political fault line: gun control and crime. Tables like the one for cream were created with randomized data, sometimes indicating that gun control decreased crime and other times suggesting it increased it. With the question now firmly politicized, the problems were randomly distributed among liberals and conservatives.

When the results were analyzed, something extraordinary emerged: mathematical ability ceased to be a predictor of how well subjects fared. Liberals were remarkably effective at solving the ratios when they suggested gun control reduced crime, but when confronted with data that suggested the opposite, their mathematical skills abandoned them and they tended to get the wrong answer. Conservatives exhibited precisely the same pattern, only in reverse: they were able to solve the problem when it suggested that lax gun laws reduce crime, but completely flunked it when the data pointed in the opposite direction. More alarmingly, one's level of skill didn't seem enough to overcome the impact of partisanship—on average, those with better mathematical skills were more likely to get the answer right when it aligned with their ideological position.

Kahan's body of research lays waste to the idea that an information deficit is the reason for disagreement on issues of science and technology, or of policy and evidence. Rather, it suggests that ideological motivations skew our very ability to reason. But why might this be the case? Kahan's theory is that people have a propensity to engage in *identity-protective cognition*: "As a way of avoiding dissonance and estrangement from valued groups, individuals subconsciously resist factual information that threatens their defining values." We do not separate our beliefs from ourselves—to some extent our beliefs define us. Accordingly, it is a major psychological imperative to protect this idea of who we are, and our relationships with those who share the same ideas and worldviews as us. We find it immensely difficult to differentiate our ideas from our sense of self—and too often, this condemns us to clutch at wrong-headed positions with dogmatic zeal, unwilling to countenance alternatives lest they threaten our very identity.

If this sounds strange, consider the consequences for someone who defies their group identity, and the staple unquestioned beliefs and assumptions inherent to that group. We have a tendency to inhabit echo chambers of opinion and ideology that reflect our own. This is extremely apparent in emotive subjects, whether religion, politics, or beliefs. In these realms, collective subscription to certain views reinforces ideas until they become unquestioned orthodoxy. Any deviation from these ideas can come at a high social and personal cost, including ostracization from the group. To question aspects of one's belief is often conflated with being treacherous to that view, and risks making one a pariah.

Curiously, cognitive dissonance seems to be somewhat selective. In a 2012 article, researchers at the University of Kent found that believers in conspiracy theories had a surprising ability to hold two mutually exclusive beliefs at once. In one study, the more participants believed that Princess Diana had faked her own death, the more they believed she was murdered. Similarly, another study found that the stronger a subject's belief that Osama bin Laden was already dead when US Special Forces raided his compound in Pakistan, the deeper their belief that he was still alive. Somehow, conspiracy theorists were able to accept some bizarre Schrödinger's Bin Laden, simultaneously existing in both an alive and dead state. The reason this caused no conflict to believers is that the specifics of the belief themselves were inconsequential—as long as there is a narrative of conspiracy, their worldview is protected.* As the researchers concluded, the "nature of conspiracy belief appears to be driven not by conspiracy theories directly supporting one another, but by broader beliefs supporting conspiracy theories in general."

The alarming reality is that people tend to believe what ideologically appeals to them, filtering out information that conflicts with their deeply held beliefs. This afflicts all of us to some degree and is something we need to be actively aware of if we are to have any chance of overcoming it. What feels to us like a rational position might not be anything of the sort—often, it could instead be an emotional decision dressed in the borrowed garb of rational thought, entangled with the very fabric of how we define ourselves. This makes us resistant to changing our minds, even when the available data urges it. As Jonathan Swift once observed, "reasoning will never make a man correct an ill opinion, which by reasoning he never acquired."

But ultimately, clinging to irrational beliefs is detrimental to us. Whether the issue is climate change, health policy, or even politics,

---

* There is good psychological evidence that acceptance of conspiracy theories is strongly correlated with a need for control. There's also an ego-drive component, with believers feeling better informed than their peers.

we need to be able to evaluate the available information critically without the distorting lens of ideology coloring our perception. While we may hold incredibly strong personal convictions, reality doesn't care one iota for what we believe. And if we persist in choosing ideology over evidence, we endanger ourselves and others.

9

THE MEMORY REMAINS

Satanic Ritual Abuse, Conformity of Memory, and
Why Our Recollections Are Often Contrived

"Memory is the treasury and guardian of all things."
— MARCUS TULLIUS CICERO

"Memory is an illusion, nothing more. It is a fire that needs
constant tending."

— RAY BRADBURY

In criminal trials, eyewitness testimony is afforded huge
importance. The recollections of those present at the scene of a
crime hold powerful sway over juries. Often, they're the smoking
gun that convicts or acquits a suspect. Yet, despite the stock we
put in such evidence, this trust may be misplaced. Aside from the
real risk of misidentification, witnesses regularly reconstruct fragments of information into a coherent but incorrect narrative. We
each have a predisposition to store information in a way that makes
sense to us. These personal schemas are shaped by our own experiences, cultural conditioning, and even prejudices. Subconsciously,
we alter our recollection of events and their order to fit these
factors, which in turns distorts our perception. This process is so
seamless we're completely unaware when it occurs. The memory
feels real, capturing the gist of our experiences, but as an objective
testament it is fundamentally flawed. The Innocence Project, a justice reform group, has found that mistaken eyewitness testimony
was a major factor in 73 percent of wrongful convictions.

This doesn't appear to be a product of deception, but rather a
glitch in how we remember. Similarly, the fact that the testimonies
of eyewitnesses often conflict stems from the same fundamental
malleability. But why? Luckily, the vagaries of memory have long

fascinated neuroscientists, and Oliver Sacks wrote in detail on the topic. In his autobiography, he recounted a terrifying ordeal with a raging thermite bomb that exploded near his childhood home during the London Blitz, almost razing it to the ground. Sometime after publication, his elder brother informed him that he had not in fact been present when the bomb landed—rather, his vivid memory had sprung from a detailed letter their eldest brother had written, which had enthralled the young Oliver. Somehow, the dramatic telling had fused with Sacks's own memories:

> It is startling to realise that some of our most cherished memories may never have happened—or may have happened to someone else. I suspect that many of my enthusiasms and impulses, which seem entirely my own, have arisen from others' suggestions, which have powerfully influenced me, consciously or unconsciously, and then been forgotten.

Sacks's experience is not unique. There is a conception of memory as a pristine recorder of all that went before: a repository of all the experiences, emotions, and events that shape who we are. But the truth is that, as real as our memories feel to us, they are at best approximations, constantly rewritten and eroded, corroding and changing with time. Our minds have a fantastic tendency to reshuffle events and alter narratives, embellishing or simplifying sequences. Far from being an infallible record of events, our memories are therefore notoriously easy to manipulate, by ourselves and others. Each of us is the unreliable narrator of our own life, and we are more suggestible than we imagine.

This fluidity of memory can be disconcerting when baldly confronted. For me, 2007 was an eventful year replete with a chain of bizarre events afflicting my immediate circle. As a vent for the chaos of that year, I kept a detailed journal through those testing months. Years later, a friend and I were reminiscing over the intensity of that time—as a writer, he was considering a fictionalized approximation

of those events as a backdrop for his story. Keen to help, I dug out
the journal for reference. What was striking was just how divergent
our memories were. To a varying extent, the journal contradicted
our recollections and chronologies, which were in turn inconsistent
even between us. Being curious, I delved into the relics of that era in
the form of old emails and messages. These confirmed the accuracy
of the written account but left an uncomfortable conclusion: Our
memories had independently restructured details and sequences
with the passing of time.

This vulnerability might surprise us, yet as eloquently stated
by memory researcher Christopher French, "all of your memories
are reconstructions, and to a greater or lesser degree there will be
distortions in there." So tenuous are the chains of memory that it's
even possible to conjure false memories. Elizabeth Loftus, a seminal
figure in the world of memory research, demonstrated this aspect
with her "Lost in the Mall" technique. In this series of experiments,
Loftus and her students investigated whether a wholly synthetic
event could be implanted and accepted as a real memory. The investigators would garner a series of vignettes from early childhood
experiences, supplied by the subject's family. Among these true
stories, in conversation they inserted a solitary false experience—a
story of a time when the subject was separated from their family
in a shopping mall and rescued by an elderly person who reunited
them with their family. Despite this event being a fabrication,
roughly 25 percent of subjects not only came to believe that it had
happened, but began embellishing with details, convinced of the
reality of this event. Nor was this a unique blip; in other memory
implantation experiments, a global average of around 37 percent of
subjects identify false memories nurtured by the researchers as real
memories, adding detail to them.

Moreover, if a sham memory is accompanied by other ostensibly
supporting pieces of evidence, the level of alleged recall and detail
rises dramatically. In one experiment, researchers provided subjects
with a doctored image of them as a child in a hot-air balloon. This

visual flourish dramatically increased the number of subjects giving detailed accounts of this nonexistent event. These of course were not deliberate attempts by the subjects to deceive the scientists, but a consequence of how we interpret and form memory. In the words of Loftus: "Memory does not work like a video recording device where you just record the event and play it back later—it's a little bit more like a Wikipedia page where you can go in there and change it—but so can other people."

The idea that our precious memories can be altered—not only by ourselves but by others—is an alarming thought, but one well supported by scientific evidence. Our memory is hugely affected by social factors. *Conformity of memory* is one such phenomenon, where an individual's report of a memory influences the recall of another. In a particularly famous example, witnesses to the murder of the Swedish Foreign Minister Anna Lindh were held in a room together while awaiting questioning. Contrary to explicit instructions, they discussed their versions of events, unduly biasing each other's memories. When the perpetrator, Mijailo Mijailović, was finally apprehended after being caught on security camera footage, his appearance did not remotely match the description provided. Despite how cohesive the eyewitness accounts seemed, the corroboration was entirely illusory.

Under conditions of high stress, too, our recollection falters. One revealing study demonstrated that highly trained soldiers under mock prisoner-of-war conditions consistently misidentified their aggressor after the exercise, with participants prone to dubious reconstruction after the event.

Yet perhaps the most ignominious failure to recognize the unreliable nature of our recall is the Freudian concept of repressed memories. In 1973, Canadian psychiatrist Lawrence Pazder took on a new patient, Michelle Smith, at his private practice in Victoria. For three years, their sessions were largely mundane. Following a miscarriage in 1976, however, Michelle suffered a bout of depression and her sessions intensified. Cryptically, she told Pazder that she had something important to tell him, but no recollection of what that was. Shortly afterward, one of their sessions devolved into Smith screaming unstoppably for 25 minutes before resuming speech in a child's voice. Intrigued by this bizarre behavior and determined to unlock Michelle's secret, Pazder opted to employ a relatively new tool in the therapist's arsenal—hypnotic regression.

The introduction of hypnosis heralded a sea change, unlocking all manner of formerly repressed memories. The story revealed was shocking. Michelle alleged that, from the age of five, her mother had offered her up to a satanic cult where she was ritually abused in the most debased ways imaginable. In that trancelike state, her seemingly unlocked memories materialized, replete with gruesome detail. Smith reported that she had been locked in a cage, and physically and sexually abused in dark ceremonies. She spoke of witnessing the murder of babies and being doused in their blood, with her ordeal culminating in an uninterrupted 81-day ritual of orgies and violence in the name of the devil. After this, the cult elders wiped her memories and invoked black magic to remove her scars.

Pazder, a devout Roman Catholic, was convinced. He became completely enthralled in Smith's therapy, spending over 600 hours regressing Smith to uncover more. His devotion eventually cost him his marriage, and soon afterward he and Smith became romantically involved. In 1980, they published the now-infamous *Michelle Remembers*, an account of Smith's testimony in which Pazder coined the term "ritual abuse." Despite the disturbing content, the book became an immediate bestseller. Religious and evangelical groups viewed the book as proof positive that satanic elements were active throughout America. Law enforcement agencies were besieged by claims of ritual crimes. As similar claims emerged all over the US, Pazder was lauded as an expert, even bringing claims of organized satanic groups to the Vatican itself.

Yet, for all the furor and media coverage surrounding *Michelle Remembers*, what was lacking was basic skepticism. Pazder had initially alleged that the Church of Satan was behind the abuse, only to withdraw this claim after a legal challenge by Anton LaVey, founder of the Church of Satan.* Worse, Michelle's claims were directly contradicted by a wealth of evidence. Smith's identity was quickly exposed and her mother revealed as Virginia Proby, who had died of cancer when Smith was a teenager. Contrary to the assertions in *Michelle Remembers*, the real Mrs. Proby had been a caring, compassionate woman. Michelle's father, Jack Proby, was able specifically to refute many of the book's allegations too, and he filed an intention to sue against the book's publishers, which then stopped a proposed movie adaptation.

Upon the most cursory of observations, Smith's vivid account crumbled. In addition, her more lurid details—such as the claim that Jesus and the Archangel Michael suppressed her memories until the "time was right"—should have marked her testimony as suspect. Instead, the accounts tapped deep into the national psyche, and a chorus of voices proclaimed that ritual abuse was rampant all over America. In a horrific error of judgment, *Michelle Remembers* was even used in some instances to train social workers. Many in law enforcement and social protection became convinced of the reality of a malevolent network of abusers, and grew hypervigilant against the odious specter of satanic ritual abuse (SRA).

Given this priming, what ensued was perhaps inevitable. Allegations of organized abuse emerged rapidly from preschools all over the country, each richer in horrifying detail than the last. In 1982, prosecutors claimed to have unearthed a ritual pedophile group in Kern County, California, with up to 60 children testifying that they had been abused and 36 people convicted. In 1984, the Fells Acres Day Care Center in Massachusetts saw a spate of convictions based on the testimony of children who had claimed, among other charges, that they had been raped with knives, despite the lack of

---

* It's important to note that, despite misconceptions, the Church of Satan does not condone abuse of any sort. LaVey Satanism doesn't even believe in a theistic notion of a devil. Rather, the church is focused on individualism, taking its name from the Hebrew term for "one who opposes."

any physical evidence, and assaulted by robots and clowns in a secret room. All across America, an explosion of trials for ritual abuse were held throughout the 1980s and on into the 1990s; in 1991, children in Oak Hill, Texas, testified that they had been sexually abused by Satanists in white robes and forced to dismember crying babies.

The incredible claims outlined by these children seemed to vindicate those who believed in SRA. In general, media coverage expressed little doubt. And in every one of these cases, convictions were obtained despite the complete lack of any physical corroborating evidence. But for many people, physical evidence wasn't deemed especially important. After all, what possible motivation could children have to lie? And surely, reasoned the believers, there was simply no way that children would experience false recall with such detailed, graphic and profoundly sexual accounts. To untangle this, we need to look at perhaps the most infamous event, and the archetype of all satanic panics: the infamous McMartin preschool trial of 1984.

The McMartin family operated a preschool in the upmarket area of Manhattan Beach, California. In 1983, Judy Johnson informed police that her son, a student at the preschool, had been molested by her estranged husband and by teacher Ray Buckey, son of school administrator Peggy McMartin Buckey and grandson of founder Virginia McMartin. Johnson also alleged that staff at the daycare center had sex with animals and that Ray Buckey could fly. The police questioned Buckey, but unsurprisingly they found him completely grounded and zero evidence supporting Johnson's claim. Even so, the investigating force sent a letter to the parents of around 200 children, suggesting that their children might have been abused. The letter, which encouraged parents to question their children, also provided specific details, suggesting that parents question their children about being tied up by Buckey.

The response was a landslide; within a few weeks, several hundred children had been interviewed by the abuse charity Children's Institute International (CII). The CII center, run under the stewardship of social worker Kee McFarlane, employed a curious technique when questioning the children. Specifically, CII interviewers made suggestions about events, inviting children to role-play and pretend, in the hope this might encourage them to be more forthcoming about events they had experienced. However laudable their intentions, their questioning was so suggestive as to be positively leading. It's worth taking a moment to look at the kind of questions the CII put to the children:

Interviewer: Can you remember the naked pictures?
Child: [Shakes head no.]
Interviewer: Can't remember that part?
Child: [Shakes head no.]
Interviewer: Why don't you think about that for a while, OK? Your memory might come back to you. . . .
Interviewer: You see all the kids in this picture? Every single kid in this picture has come here and talked to us. Isn't that amazing? . . . These kids came to visit us and we found out they know a lot of yucky old secrets from that old school. And they all came and told us the secrets. And they're helping us figure out this whole puzzle of what used to go on in that place. . . .
Interviewer: How about Naked Movie Star? You guys remember that game?*
Child: No.
Interviewer: Everybody remembered that game. Let's see if we can figure it out.

---

* This later transpired to have arisen from a Californian playground taunt: "What you say is what you are / you're a naked movie star!"—completely devoid of any gruesome origin.

In hindsight, the issue is incredibly obvious. In effect, they led the young witnesses toward a false testimony, either by rejecting unambiguous answers from the children or introducing new ideas to them. Young children, suggestible and eager to please, tried hard to answer the questions posed by the adults with the response they perceived the adult to want. And worse, such suggestion even triggered terrifying memories of abuse in children who had never experienced such events. Such a technique was deeply flawed, and children subsequently were coached into ever-stronger claims. The hallmarks of all this went undetected by overzealous believers and an uncritical media, despite increasingly outlandish assertions from the children. These claims included stories of secret underground tunnels where abuse occurred, of flying witches. Another embellishment—that of frequent journeys in hot-air balloons—could have been taken from the work of Loftus herself. One of the children even identified the actor Chuck Norris as an abuser from a lineup of photographs.

In spring 1984, Virginia McMartin, Peggy McMartin Buckey, and Ray Buckey were charged alongside a number of other staff members with 321 counts of child abuse involving 48 children. In the almost two years of hearings that ensued, the prosecution, led by attorney Lael Rubin, outlined a shocking narrative of SRA. By this time considered senior figures in investigating allegations of satanic abuse, Michelle Smith and Lawrence Pazder met with the children involved, helping to shape their testimonies. Media coverage was intensely skewed toward the prosecution and, despite the lack of any physical evidence, all were held for trial. By 1986, the initial complainant, Judy Johnson, who had long suffered from schizophrenia, succumbed to alcoholism. The children's testimony was contradictory and evidently coaxed, prompting the district attorney to slam the case as "incredibly weak." He moved to drop charges against all defendants bar Ray and Peggy McMartin Buckey. By 1990, charges had been dropped against the Buckeys, who had spent more than five years in jail without conviction.

The trial had taken seven years and cost over $15 million with no convictions, rendering it one of the most expensive trials in American history. In the aftermath, the school was deemed too tainted to remain, and was demolished. No evidence of secret underground tunnels was ever found. As the child witnesses grew up, some recanted their testimony, commenting that pressure by their interviewers had shaped their recall. A review of the videotaped testimony by a British psychologist, Michael Maloney, damned the questioning technique as coercive and directive, concluding that "many of the kids' statements in the interviews were generated by the examiner" rather than by the children themselves. The true abuse these children suffered was not satanic in origin, but the false memories implanted in them through questionable techniques.

The only positive that has come out of this awful debacle is that it prompted a review of how young witnesses should be questioned, acknowledging that it is all too easy to implant false recollection, even with the best of intentions. For all the stock once placed in *Michelle Remembers*, the veracity of hypnotic recall has now been completely undermined, with copious evidence demonstrating that memories extracted by this technique are likely to be nothing more than fictions. As incredibly unfair as the treatment of the McMartin staff was, they were luckier in many respects than others who were swept up in the moral panic over SRA. In the deluge of similar trials, hundreds of others were convicted on equally flawed accounts in the 1980s and 1990s. While many of these convictions have since been quashed, at the time of writing there are still scores languishing in jail for utterly fictitious crimes.

None of this is to detract from the sheer utility of memory, or to imply that the common distortions we all experience are akin to deliberate falsehoods. Our tendency to confabulation is a seemingly inescapable part of the human experience. While this is unavoidable, it does not, a priori, mean one's account of events is amiss—but it does mean that it is a distinct possibility. Unavoidably, this thorny aspect often comes up in trials concerned with inter-personal crimes such as abuse and sexual assault, where the only evidence may be witness accounts and there are conflicting versions. In many cases this conflict will be the result of dishonesty by one of the parties, but the uncomfortable truth might be that in some cases the recall of one or more parties is simply inaccurate or externally influenced. This facet means that deciding the sequence of events in such cases is fraught with difficulty, even if no party has an intention to deceive.

Unsurprisingly, this can trigger passionate reactions; we are so deeply wedded to the simple dichotomy between truthful accounts and outright lies that we neglect the gray haze of dubious memory. Memory expert Elizabeth Loftus herself has appeared as an expert witness in several high-profile criminal trials, outlining the ethereal nature of memory. Her informed testimony has undoubtedly prevented a great many miscarriages of justice, but this has not endeared her to some. Consequently, she has been threatened legally and with violence throughout her career. For this, Loftus has been honored with Sense About Science's Maddox Prize for standing up for science in the face of adversity, an acknowledgment of the abuse she has endured for presenting the scientific case. Despite these threats and actions, Loftus and other memory researchers like her have persevered and shown that our minds all too readily can drastically distort our memories and weave the darkest of fictions with just an inkling of suggestion.*

Regression and other suggestive methods have been completely disgraced as techniques for uncovering memory and have instead

---

* Importantly, there is no intention to deceive with false memories. Loftus herself has recounted her own experiences with false memories; when she was a teenager, her mother drowned. Years afterwards, a relative told Loftus she had found the body, which later triggered a series of painful memories of the scene. Yet Loftus was in time able to ascertain with certainty that she had not made the discovery, and her relative came to realize he had been mistaken. Her memories here were false, a testament to the power of suggestion.

been shown merely to distill false narratives. Even so, hypnotic regression remains dangerously popular in some quarters, and is still used despite its damaging legacy. Today, clumsy application of these long-debunked tropes causes serious and needless rifts in families. As British psychologist Christopher French states:

> I have three wonderful daughters—two teenagers and one young adult. I can hardly imagine anything more horrible than the prospect that one of them might one day enter therapy for help with some common psychological problem such as anxiety, insomnia or depression and, at the end of that process, accuse me of childhood sexual abuse on the basis of "recovered" memories. Even though I would know with absolute certainty that such allegations were untrue, the chances are that nothing I could say or do would convince my accusers of this.

This is not merely hypothetical: There exist support groups around the world specifically for victims of false memories. But on a more mundane level, even without the theater of regression, our memories can lead us astray. Crucially, we cannot overlook the power of suggestion on our malleable minds. The idea that our recollection can be manipulated is disquieting. This conjures up the sinister editing of historical records described in George Orwell's *1984*:

> Day by day and almost minute by minute the past was brought up to date. In this way every prediction made by the Party could be shown by documentary evidence to have been correct, nor was any item of news, or any expression of opinion, which conflicted with the needs of the moment, ever allowed to remain on record. All history was a palimpsest, scraped clean and reinscribed exactly as often as was necessary. In no case would it have been possible, once the deed was done, to prove that any falsification had taken place.

The idea of having our mental records edited by some outside agent remains every bit as appalling. And certainly, we might have grounds to be concerned about the influence of media and our own social groups on our memory and perception. The powers of media attention, social pressure, and implicit suggestion are often enough to sway our recall and perception. But perhaps an even more appalling thought is that, even without external influence, our memories are as imperfect a record as the newspapers of Orwell's novel, strewn with error and outright invention. Memory is vital to our very being, but we must be wary of treating it as infallible. To do so is to anchor ourselves to an unstable rock, and risk being sunk.

10

DAGGERS OF THE MIND

Ghosts, Demons, Apophenia, Pareidolia,
and Other Ways Our Senses Fail Us

Our minds are instruments of computation and reflection, and our senses the gatekeepers of all we perceive. We are besieged by a constant maelstrom of sounds, visions, tastes, smells, and sensations that our brain effortlessly unscrambles. Without any conscious thought, we readily distinguish the sonic texture of birdsong from crackling flames; the slightest touch discriminates between cold steel and rugged oak, and a glance differentiates ocean waves from billowing clouds. Our senses are the inputs from which we compute the world, efficiently sifting through life's symphony of signal and noise.

But is our unwavering trust in our senses misplaced? That our senses can deceive us is hardly surprising, and long known—Macbeth's famous soliloquy pivots around the apparition of a dagger, and the eponymous villain musing on the nature of that which he perceives:

> Is this a dagger which I see before me,
> The handle toward my hand? Come, let me clutch thee.
> I have thee not, and yet I see thee still.
> Art thou not, fatal vision, sensible
> To feeling as to sight? Or art thou but
> A dagger of the mind, a false creation,
> Proceeding from the heat-oppressed brain?
> I see thee yet, in form as palpable
> As this which now I draw.

The equivocal nature of our perceptions was a frequent theme in Shakespeare, who left the true nature of the ghosts and visions in his plays deliberately ambiguous. But while Shakespeare employed the paranormal as a dramatic device and a reflection on the unreliable nature of our senses, belief in the supernatural has long been strongly held. Accounts of paranormal phenomena are pervasive throughout history, from accounts of alien abduction to spiritual encounters with deities. More frequently, the idea that the dead can communicate with and influence the living long after their passing is an enduring belief—accounts of hauntings, visions, and ethereal messages transcend geography and culture, and are found across all spheres of human life.

The sheer ubiquity of these accounts is enough to give one pause; while some of these tales must be outright fabrications or symptomatic of delusional disorders, this cannot in isolation account for all such happenings. Many of us are familiar with sincere personal anecdotes about an inexplicable encounter with spirits, or perhaps have had such an experience ourselves. While we're aware that human perception is fallible, the extraordinary level of detail in these accounts from sober individuals tends to erode our doubt. Even if our senses are imperfect, how could our minds mislead us in such specific and convincing ways? To resolve this, we need to understand more about our perception to see why it can be led astray.

Part of the answer is that we are adept at finding patterns, even when none exist. At the height of the spiritualist trend, the popular magazine *Scientific American* asked Thomas Edison whether his inventions might be used to communicate with the dead. Edison, inventor of the phonograph and inveterate self-promoter, replied that if spirits were capable of influence, then electric recording apparatus would better detect them than anything else available. Although the fetishism for spiritualism declined in the early twentieth century, the idea never fully faded. Following the advent of reel-to-reel taping in the 1950s, the concept underwent a renaissance when photographer Attila von Szalay began recording alleged instances of the dead communicating via tape, in what was christened Electronic Voice Phenomena (EVP).

Interest in EVP spilled over into the cultural mainstream, remaining a staple of paranormal enthusiasts to this day. But while believers interpret the static cacophony as messages from the afterlife, the reality is that there is nothing in the apparent messages that cannot be explained by interference and wishful thinking. The psychological rationale for this is *apophenia*, or the perception of patterns in random data. In a similar vein, the advent of home recorders led to tales of hidden messages in popular music when played backward. This prompted the curious spectacle of two disparate populations obsessed with playing records backward: teenage music obsessives with copious free time, and morally outraged denizens of the religious right convinced of devilish codes embedded in rock.

This was a misguided (if amusing) endeavor. It is incredibly difficult to render a series of sounds so that they convey different meanings in reverse. Even so, legends arose about sinister stories hidden on records; Led Zeppelin's "Stairway to Heaven," when played backward, allegedly tells a dark tale of demonic abuse including the words "there was a tool shed, where he made us suffer, sad Satan." Queen's "Another One Bites the Dust" allegedly contains a hidden message emphasizing the joys of cannabis.* Such is the power of apophenia that ostensible messages can be made out, given enough suggestion.

While much of the moral panic over hidden messages is little more than a bizarre historical footnote, it bears remembering that its impact at the time transcended the more paranoid elements of the evangelist communities. Perhaps nothing better typifies the prevalence of such beliefs than the curious trial of Judas Priest. As premier exponents of British metal, Judas Priest had a string of huge-selling albums to their credit. Among these was their 1978 album *Stained Class*, containing the single "Better by You, Better

---

* As a hard-rock-loving music snob, I spent an unhealthy portion of my teenage days reversing audio files to catch glimpses of these alleged messages. The result was inevitably gibberish. Never let it be said that teenagers can't make their own entertainment.

than Me"—a song that was to become a thorn in the band's side. Years later and half a world away, two Nevada teenagers, James Vance and Ray Belknap, were listening to the album one evening in 1985, in the sweltering desert heat. Inexplicably, the two young men put guns to their heads and pulled the triggers.

Belknap was killed instantly, but Vance survived for three years before dying of drug complications. In the wake of such tragedy, the boys' families grasped for an underlying reason for such a senseless loss of young life, a simple scapegoat for the complexity of suicide. Attention turned to the pair's fondness for heavy metal, with Judas Priest identified as potential catalysts for these deaths. However, nowhere in the band's considerable output could one find anything even indirectly encouraging suicide. Still, the school's guidance teacher claimed that Vance blamed the song for triggering events. The family and their prosecution counsel alleged that the song contained a hidden message in reverse, densely masked in the mix—an ominous, cryptic command prompting the suicide pact: "Do it."

This was news to the band themselves, who adamantly maintained there was no such message there. Vance's family insisted, nonetheless, that they could clearly ascertain it, and that this constituted a subliminal order for the boys to end their lives. Despite misgivings, Justice Jerry Carr Whitehead ruled there was a case to answer. Compounding matters, he further decreed that any subliminal messages would be exempt from first amendment protection. The case was ultimately dismissed in August 1990. In a write-up for *Skeptical Inquirer*, psychologist and witness for the defense Timothy E. Moore explained precisely why the utterance was heard and, more importantly, how easily one could subconsciously be manipulated into hearing such ghostly imperatives:

> Perception is an active, constructive process. Consequently, people often see or hear what they are predisposed (or encouraged) to perceive. A diligent search entailing the isolation and

amplification of dozens of snippets from a three-minute heavy metal rock recording would probably yield some intelligible words or phrases that would not be intelligible under normal listening conditions. In fact, it would be surprising if a few such "discoveries" were not made.

Apophenia is not confined to auditory phenomena; it can flummox any of our senses, skewing our perception. As vision is of such fundamental importance in how we interpret the world, it's not surprising that we're particularly vulnerable to perceptual illusion. *Pareidolia* is the psychological phenomenon of perceiving a known pattern in a stimulus when none in fact exists. As innately social creatures, we're predisposed to seeing faces and figures. Further, most of us (with the possible exception of those high on the autism spectrum) can assign "emotions" to these faces. This may be patently absurd; yet so important is this to us as a species that we do it instinctively. Pareidolia has long been an important artistic device. In the famous 1566 painting "The Jurist" by Giuseppe Arcimboldo, the "face" in the painting is in fact composed of fish and chicken. Leonardo da Vinci too encouraged artists to make use of this perceptual quirk to better their work, advising how one could gaze at stone walls to see:

Resemblances to a number of landscapes, adorned with mountains, rivers, rocks, trees, great plains, valleys, and hills, in various ways. Also you can see various battles, and lively postures of strange figures, expressions on faces, costumes and an infinite number of things . . .

Leonardo was fully aware as he gazed upon fifteenth-century masonry that this was a canvas for his powerful imagination, just as we today seek out figures in the clouds on a summer's day for our own amusement. But pareidolia isn't always conscious, nor obvious. When NASA's *Viking 1* reached Mars in 1976, one photo it

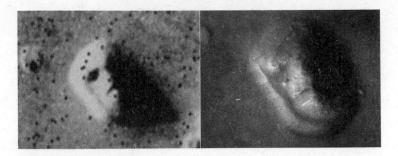

Left: The "face" on Mars captured by *Viking 1* in 1976. *Right:* A high resolution image of the same area by the Mars Reconnaissance Orbiter, showing the ostensible face to clearly be a rocky mesa. (© NASA)

sent back appeared to depict a vague humanoid face in the region of Cydonia. This led to frenzied speculation in some quarters that it might be the remnant of an ancient alien temple, proof of an extinct Martian civilization. However, the "face on Mars" was an illusion, a product of Viking's low-resolution imaging, with subsequent high resolution imaging clearly showing it to be a mesa. This of course has been no impediment to the army of conspiracy theorists who still insist that NASA covers up evidence of alien life.

For the devoutly religious, random patterns may be heralded as minor miracles. Listing the objects on which the faces of religious figures have appeared would take the rest of this book, but a non-comprehensive list includes wood stains, candlewax, and spaghetti with the face of Jesus, a tree-callus monkey god, dust spelling the name of Allah, and even a grilled cheese sandwich with the face of the Virgin Mary.

But it is instances of apparently paranormal phenomena where unchecked perceptual faults can steer us falsely. Since the dawn of human storytelling, we have had tales of ghosts, demons, and aliens. Belief in these phenomena remains strong even centuries after the Enlightenment. A 2017 YouGov survey found that 50 percent of respondents believed in ghosts, and a 2015 Pew survey found 18 percent claimed to have personally encountered them. It would be churlish to dismiss all these accounts as hoaxes, and the sheer abundance of anecdote and personal experience is often enough to give the skeptical pause.

Nor do encounters with ghosts seem to be the preserve of the skittish; following the Second World War, the formidable Winston Churchill was visiting the White House. After a long soak in the tub, replete with Scotch and cigars, Churchill staggered into the adjoining room only to be confronted by the ghost of Abraham Lincoln. Unflappable even when spectacularly naked, the former prime minister remained droll, muttering, "Good evening, Mr President; you seem to have me at a disadvantage," before Lincoln's grinning specter dissipated. A cynic might remark that this encounter was more a product of the copious quantity of Scotch imbibed, or Churchill's talent for perpetuating his own legend with embellished tales. Whatever the veracity of this account, a significant proportion of the population maintain that they have come across a spirit of the deceased.

Often, these experiences take place close to bereavement. When I was 17, an uncle with whom I was very close died unexpectedly. He had been such a fixture in our house that the guest room was often simply referred to as "Michael's room." One evening not long after his death, out of the corner of my eye I spotted a shadow climbing the stairs to the room. Though I knew rationally it couldn't be, my mind still leapt to association with Michael, compounded by a longing to talk with him. Such tales are not rare; the experience of catching "shadow people" (also known as "black mass") from the periphery of one's vision is a universal experience. Much of this is explained by simple pareidolia, yet not everyone is equally affected—believers in the paranormal are much more likely to mistake random patterns for intentional forces. In one experiment, believers saw a "walking" figure in a pattern of random lights far more than their skeptical colleagues. Other brain-imaging experiments with random moving shapes have shown that those inclined to supernatural beliefs show more brain activity associated with intention, ascribing motives to the mess of random motion onscreen.

While this provides part of the explanation for the enduring prevalence of supernatural belief in human society, a believer may well counter that this does not address the feeling of a presence common to many narratives of haunting. Frequently, a sensation of presence from a disembodied specter is the quality that convinces one that the experience is something beyond mere misidentification. This presence too is often frightening and malignant, and the venue for these encounters is commonly a place of some foreboding or ambiguity; graveyards, lonely houses, and darkened cellars feature heavily in Gothic literature for good reason. In popular telling, those who stumble upon ghosts not only see an ethereal apparition, but also experience an unmistakable feeling of presence and unreconstructed dread.

Such accounts are manifold in fiction and first-hand experience. But, gripping as these tales are, there is a decidedly more earthbound explanation for the enduring *feeling of presence* (FoP): our brains struggling to make sense of conflicting signals. FoP is often a symptom of schizophrenia or epilepsy. Research suggests that these sensations might be related to sensorimotor mismatch, or a disconnect between what we perceive and what our bodies feel. Evidence suggests that damage to the frontoparietal cortex is associated with this syndrome. This region is involved in awareness of self and integration of the internal and external stimuli we encounter. Lesions here can induce disturbing sensations of presences following the individual. This effect can be simulated in perfectly healthy volunteers. In one experiment, a robot was designed to sit behind a blindfolded subject, mimicking their exact movements. Under normal operation, subjects reported a feeling that they were touching their own backs, an unusual but distinctly non-hallucinatory sensation. But with the introduction of a time delay as small as a half-second, subjects became disturbed, reporting sensations of presence. So disturbing was this illusion that many subjects asked the experimenters to stop testing.

Those undergoing extreme events are prone to experiencing an alien presence. The "third man" factor was first recounted by explorer Sir Ernest Shackleton after he and his party felt the presence of an incorporeal agent on the final leg of their Antarctic quest. This is particularly common with mountaineers, extreme-marathon runners, shipwreck survivors, and solo sailors. On a solo jaunt up Mount Everest, British mountaineer Frank Smythe was so convinced by this hallucination of an invisible climbing partner that he broke off a piece of cake and offered it to the spirit climber. Third-man syndrome appears to arise from conditions of monotony and isolation, correlated with darkness and barren landscapes. Severe cold, injury, hunger, and thirst are also associated with the experience. Extreme fatigue can also manifest as visions of a ghostly presence, with sleep deprivation confounding our senses and resultant perceptions. Aviation pioneer Charles Lindbergh recounted the following lucid hallucination on a flight to Paris:

> The fuselage behind me becomes filled with ghostly presences—vaguely outlined forms, transparent, moving, riding weightless with me in the plane ... conversing and advising on my flight, discussing problems of my navigation, reassuring me, giving me messages of importance unattainable in ordinary life.

Because our brains are hives of electrical and chemical signaling, we are vulnerable to distortions of perception through electrical or chemical disruptions. Hallucinations are one such class of events, with a spectrum of causes. Hypnagogic hallucinations are the auditory and visual perceptions that can occur when one is falling asleep or waking up. Fleeting images, snatches of disembodied speech, and even tactile sensations while on the brink of sleep are common, and those experiencing them are generally aware that the stimulus is illusory. Edgar Allan Poe wrote of the fancies he received "only when . . . on the brink of sleep, with the consciousness that I am so."

Less pleasantly, the phenomenon of sleep paralysis is often accompanied by darker hallucinations. During rapid eye movement (REM) sleep, our bodies undergo a muscle arrest known as atonia to prevent us from acting out our dreams. But atonia can go askew; sleepwalking is one result. Other parasomnias include sleep sex (sexsomnia), where an individual engages in sexual activity while asleep. Less common are cases of homicidal sleep-walking, where a person acting in a dream-state kills another, typically family members. One fascinating exception involved French detective Robert Ledru, who in the 1880s was investigating a murder on a beach in Le Havre. The killer had left footprints with a missing a big toe. Seeing this, Ledru turned himself in for sleepmurder. Skeptical fellow officers were reluctant to buy this story, until they tested it by observing him sleepwalking—and using a gun they had left in his cell during this state. He was acquitted of murder but exiled to a country farm under court order to sleep alone in a locked room. This, all considered, was abundantly more sensible than leaving a suspected murderer with an active firearm.

Sleep paralysis occurs when an individual begins to return to consciousness while still paralysed by atonia. Despite being conscious, the sufferer is unable to move or speak. The experience generally lasts between seconds and minutes and is usually accompanied by a terrifying sense of a malign intruder. For some, this experience is compounded by waking hallucinations, often a demonic voice, and a crushing sensation in their chest, as if the invader is lying on top of them. As an intermittent long-time sufferer of sleep paralysis, I can vouch that this is a deeply unpleasant thing to endure, even forearmed with full knowledge of what is transpiring.

Sleep paralysis is likely responsible for one of the oldest and most universal demonic legends—Incubi and Succubi. These odious beings mount sleeping people, taking sexual advantage. Victims are unable to move, crushed under the weight of the intruder. Stories of these deviant fiends span history and the globe; incubi lurk everywhere from a Sumerian manuscript dated around 2400 BCE to the Book of Genesis and the mare of German folklore who rides the chest of sleepers, bestowing bad dreams (whence we derive the term "nightmare"). The archetype of the incubus exists the world over, from the South African Tokolosh to the Turkish Karabasan. Without knowledge of the perception-bending effects of sleep paralysis, it is completely understandable that this has historically been perceived as a paranormal attack in line with one's cultural understanding, whether this includes incubi or extraterrestrials. Historical belief in such creatures is perhaps testament to the fact we tend to place extreme stock in our perception, implicitly considering our senses to be unerring when this simply is not the case.

Chemically altering perception has been observed since antiquity, with hallucinogenic drugs eliciting seemingly real but frequently surreal visions. In Native American culture, use of psychoactive peyote was used to induce visions, which were interpreted as spiritual communication. In the Huichol religion, peyote is even a principal deity. Accidental ingestion of hallucinogenic substances is far more frightening. Ergot poisoning occurs after consumption of rye infested with the fungus *Claviceps purpurea*; its ghastly symptoms including convulsions and terrifying psychosis. This was common in the Middle Ages, where it was referred to as St. Anthony's Fire in reference both to the burning sensation produced and the order of monks dedicated to treating sufferers. Modern lysergic acid diethylamide (LSD) is derived from this very fungus, and its psychedelic effects should be sufficiently well known to be taken as real.

Sudden withdrawal of chemical agents can also alter perception. Delirium tremens (DTs) is the rapid onset of confusion caused by

alcohol withdrawal as the body tries valiantly to regain homeosta-
sis, or a stable state. Frequently those in the throes of the chemical
cascade induced by DTs are subject to terrifying auditory and
visual hallucinations, so much so that "blue horrors" has become
a euphemism for the experience. Methamphetamine addicts often
experience terrifying glimpses of shadow people in the periphery
of their vision. This typically occurs in the prolonged sleep deprivation of addiction and recovery. Like third man syndrome, the sleep
deprivation associated with amphetamine abuse seems enough to
convince even the most level-headed of a lurking presence.

Perception is malleable rather than absolute, readily perturbed
by outside influence. But more than this, our preconceived
notions shape it, and even influence our physical response to
stimuli, unbeknownst to our conscious mind. The explosion of
interest in spiritualism from the middle of the nineteenth century pivoted on the central tenet that the dead had both ability
and inclination to speak with the living. Such communication
was rarely direct, of course; across high society, mediums were
much in demand, armed with a slew of esoteric techniques for
communicating with the departed, plus a considerable flair for
showmanship. One of the most dramatic demonstrations of the
influence of spirits was "table-turning," a form of seance where
participants would sit around a table that tilted to a preset
alphabet, spelling out ghostly messages. To believers this was
proof positive of the endurance of the spirit after death. They
even posited a mechanism of action: the ectenic force, born of
ectoplasm and spirit energy.

Yet not everyone was convinced. Michael Faraday, the pioneering British physicist, was particularly skeptical about the ectenic
force. An avid experimentalist, Faraday set about testing whether
table-turning had any basis, carefully eliminating variables and
exploring alternative explanations. Using wood and rubber to
increase resistance to movement, he observed no effect on the

table's motion. Further examination led him to conclude that—far from being some bizarre supernatural event—table-turning was nothing more than "a quasi-involuntary muscular action." The ethereal motion of the tables didn't require some supernatural cause; the age-old propensity for men and women to delude themselves was quite enough.

Nor was Faraday alone in his efforts and frustrations. Meticulous experiments by French chemist Michel Eugène Chevreul also put nails in the proverbial coffin of spiritualist beliefs. Chevreul's accomplishments were manifold and for his contributions to science his name is one of the 72 inscribed on the Eiffel Tower. A pioneering scientific thinker who was, like Faraday, steadfastly opposed to charlatanism, by the mid-1800s his attention turned to a trinity of spiritualist tools: table-turning, divining rods, and magic pendulums. In an 1854 paper, Chevreul elucidated how involuntary and subconscious muscle reactions are the cause of ostensibly magic movements. Once the person holding the rod was made aware of this reaction, the movements ceased and could not be reproduced. That same year, physician William Carpenter introduced the term "ideomotor response" to describe this very phenomenon.

Ideomotor reflex underpins another perennial spiritualist parlor trick: automatic writing, the supposedly supernatural "channeling" of writing from a remote source. In automatic writing, the medium in a trancelike state inscribes ghostly messages, allegedly from beyond the grave. While this feat of disembodied writing mesmerized the British upper classes, Charles Arthur Mercier was not among them. Mercier was a psychiatrist with precious little time for nonsense, dedicating a great deal of time to debunking trance mediums. Turning his attention to automatic writing, he demonstrated that the only curious phenomenon at play was a variant of the ideomotor effect. Writing up his findings for the British Medical Journal in 1894, he comprehensively dismantled the spiritualist presumption that spirits were the cause of automatic writing, stating bluntly that "there is no need nor room for the agency of spirits, and the invocation of such agency is the sign of a mind not merely unscientific, but uninformed."

The unconscious muscle movement of ideomotor reflex demonstrated that there was no mystery underpinning the eerie happenings of late nineteenth- and early twentieth-century seances—merely a heady mixture of delusion and, frequently, outright fraud. While ideomotor reflex is an honest mistake, the prevalence of fraud cannot be understated; popular trance mediums of the era employed unadulterated showmanship in a bid to improve audience numbers, with the most outlandish gimmicks drawing the largest crowds. Medium Mina Crandon is an infamous example, performing nude seances where she secreted "ectoplasm" from her vagina. As her fame reached its zenith, she was exposed as a fraud by magician and escape artist Harry Houdini. Houdini himself had something of a passion for debunking fraudulent mediums. He viewed them as exploitive and used his expertise in sleight of hand to expose trickery, attending seances undercover to reveal fraud. So dedicated was Houdini to this cause, he was even made a judge on a *Scientific American* committee that offered a cash prize to anyone who could demonstrate supernatural powers. To this day, that prize remains resolutely unclaimed.

Perhaps unsurprisingly, Houdini's zeal for exposing frauds made him profoundly unpopular with spiritualists and cost him at least one friendship—that of Sir Arthur Conan Doyle, creator of the intrepid detective Sherlock Holmes. Doyle was a committed believer in all things spiritual, especially after the loss of his son in the First World War. To Houdini's immense frustration, Doyle remained adamant in his beliefs even as Houdini debunked them, driving a wedge between the two men. Even after Houdini's early death, Doyle steadfastly insisted Houdini himself had supernatural powers and grumbled that his unwillingness to contact the living was a manifestation of Houdini's stubbornness.

The sheer weight of combined scientific rebuttals should in theory have sounded the death knell for everything from dowsing to Ouija boards, yet they remain a stubborn part of our cultural conscience. And though the existence of the ideomotor effect has been well known for almost two centuries, the infinite human capacity for reinvention and our seeming inability to learn from our mistakes mean that we can still fall prey to the same illusions in different packaging.

In an especially ignoble example, 2013 saw businessman Jim McCormick convicted of selling useless bomb-dowsing kits to the Iraqi army—a completely unethical modern twist on the equally useless divining rod. Another inglorious 2013 entry in the ideomotor catalogue was the C-Fast, a mechanical dowsing rod sold as a quick detection kit for liver disease. These claims were eviscerated by Síle Lane of Sense About Science as "pushing hope and nothing more."

But as contemptuous as these examples are, there is a far more tragic offspring of these tired delusions: the concept of Facilitated Communication (FC). In FC, a facilitator helps move the arm of a patient to a screen or keyboard so that they may apparently communicate. For the families of the profoundly communication-impaired, the vaguest hint of such a Rosetta stone was heralded as a breakthrough—even if the evidence for efficacy was sorely missing. By the late 1980s, hope had triumphed comprehensively over experience—wondrous stories of FC "unlocking" nonverbal autistics and the severely intellectually disabled became commonplace, with anecdote traded as evidence. Overnight, profoundly mentally handicapped children became poets and savants, in some instances even publishing books with their facilitators.

But despite the enthusiasm with which FC was adopted, the warning signs of pseudoscience were clear from the outset; by as early as 1991, over 40 empirical studies showed no evidence of efficacy but plenty of evidence of facilitator input, the hallmark of the ideomotor reflex. Facilitators were not freeing the thoughts of the oppressed—they were projecting their own. As with the satanic abuse panic, it was inevitable that these narratives would take a dark turn; facilitators began to reveal messages of deplorable abuse from their disabled victims. With depressing predictability, a number of arrests based on this testimony took place. Media coverage fixated on the case of 16-year-old nonverbal autistic Betsy Wheaton, who ostensibly communicated through her assistant Janyce Boynton that her father "makes me hold his penissss [sic]." Soon after, she began detailing the most horrific abuse allegedly perpetrated by a wide cross-section of her family.

Resolving to determine the veracity of the claims, speech pathologist Howard Shane and psychologist Douglas Howler put together a simple but ingenious test: an apparatus that displayed pictures to both Betsy and her facilitator, where Betsy was asked to identify the object she saw. Unbeknownst to Boynton, in some trials, she and Betsy were deliberately shown different pictures. In every instance, Betsy communicated only what Janyce saw—incontrovertible evidence the communication came not from Betsy but from Boynton.* This lamentable situation and the ensuing backlash should have been the end for FC but, as with many pseudosciences that offer unrealistic promise, it has simply refused to die. Howler himself, a one-time believer, was acutely aware that not even such robust evidence could shift the convictions of communicators: "We had overwhelming evidence for facilitator control. It began to dawn on us that the impact on facilitators was going to be traumatic. FC had become an essential part of their belief system, an essential part of their personality."

Howler's warning has proved to be remarkably prescient: the heady promises of FC still hold sway with desperate parents, convinced they're communicating with their child. For many, a delusion of communication is preferable to a sad reality that their child may not have the requisite cognitive function to be communicative, and so FC continues to enjoy support in many quarters. A recent review on the topic stated this with weary brevity: "It is likely that

---

* In Boynton's defense, she later disavowed FC and accepted the scientific findings.

FC will continue to reinforce the assumptions of efficacy among parents and practitioners."

FC may have no more scientific credibility than a Ouija board or a dowsing rod, yet its specter haunts us even now, as it crops up with tedious regularity. Perhaps most disturbing and sad is the 2015 case of academic Anna Stubblefield, who practiced the technique on a severely disabled patient, referred to only as D.J. in court transcripts. She rapidly became convinced that, far being mentally stunted, D.J. was a savant—and more than that, a man who was confessing his unwavering love for her. In a grim sequence of events, her devout belief led to a sexual relationship with a man completely incapable of consent. Far from being a savant, D.J. had the mental capacity of a toddler. Tragically, despite her conviction for assault, Stubblefield remained steadfast in her beliefs, unable to countenance the reality that D.J.'s verbose expressions of love were nothing more than projected fantasy, authored solely by her own subconscious.

There is a crucial point to all this. We tend to place a heavy emphasis on our personal experience, often to the exclusion of other possibilities. Yet the stark reality is that neither our memories nor our perceptions are always trustworthy. Even with the best of intentions, we are unreliable narrators of our own experiences. As we've seen previously, we place great stock in personal stories—but though we may have no intention to mislead, the reality of flawed perceptions means we must rule out alternative explanations first. The truth is that no one's account is immune to subversion by the influence of the world around us—including our own.

11

GREAT EXPECTATIONS

Astrology, Electromagnetic Hypersensitivity, Placebos,
Nocebos, and the Dunning-Kruger Phenomenon

If you'll grant me the indulgence, let's undertake a small experiment. Take a glance at the following statements and assess how well these apply to you. For convenience, you might give each trait a mark for accuracy between 0 (lowest) and 5 (highest).

1. You have a great need for other people to like and admire you.
2. You have a tendency to be critical of yourself.
3. You have a great deal of unused capacity, which you have not turned to your advantage.
4. While you have some personality weaknesses, you are generally able to compensate for them.
5. Your sexual adjustment has presented problems for you.
6. Disciplined and self-controlled outside, you tend to be worrisome and insecure inside.
7. At times you have serious doubts as to whether you have made the right decision or done the right thing.
8. You prefer a certain amount of change and variety and become dissatisfied when hemmed in by restrictions and limitations.
9. You pride yourself as an independent thinker and do not accept others' statements without satisfactory proof.
10. You have found it unwise to be too frank in revealing yourself to others.
11. At times you are extroverted, affable, sociable, while at other times you are introverted, wary, reserved.
12. Some of your aspirations tend to be pretty unrealistic.
13. Security is one of your major goals in life.

How well did this assessment apply to you? If it seems uncannily accurate, you may now be concerned you've been targeted by some Machiavellian market research project. Rest assured; any appearance of insight is entirely illusory. What you've just undertaken has a rather more curious historical origin. In 1948, psychologist Bertram Forer compiled personality assessments for his students. The students read these in private and, just as you have done, they were asked to rate their assessment out of 5 for accuracy. Forer's students were impressed with his seemingly uncanny knowledge, awarding it an average mark of 4.26.

Unbeknownst to the students, however, Forer had given the exact same "analysis" to everyone, composed totally of lines pilfered indiscriminately from various horoscopes–the very assessment you have just read. This experiment was the first academic demonstration of the Forer effect, the observation that people tend to give high ratings for personality descriptions they believe are specific to them, even though they are sufficiently vague to apply to a great multitude of people. These kinds of open-ended statements are known as Barnum statements, after the legendary circus promoter and hoaxer P. T. Barnum.*

Subsequent investigation revealed further nuance–people are far more likely to be fooled by this approach if they believe in the authority of their assessor and if they believe the reading is specific to them. Most tellingly, perhaps, the content of the assessment matters too, with subjects far more likely to believe flattering statements over more negative ones. One especially revealing experiment involved researchers writing accurate personality assessments for subjects, coupled with a generic Barnum statement. When asked to pick the most accurate description, the majority of

---

* Barnum is often associated with the phrase "there's a sucker born every minute," but there's no evidence he uttered these words. Still, Barnum's penchant for deceiving audiences remains the stuff of legend. My favorite anecdote involves his problem with customers lingering around exhibits, which he rectified by erecting signs reading "This way to the egress." Excited spectators, unaware of what this word meant, would rush toward it–only to find themselves outside. Adding insult to injury, Barnum had his staff charge them again if they wished to return.

subjects opted for the fawning Barnum statement over the more personalized assessment—the triumph of vanity over reality.

This aspect of the Forer effect goes a long way toward explaining our collective propensity to cling to long-discredited ideas. Astrology—the belief that the movements of planets and stars influence human fate—has a long heritage of criticism in both ancient and modern eras. As far back as the twelfth century, philosopher and physician Maimonides dismissed the concept unreservedly, wearily declaring, "astrology is a disease, not a science"—a sentiment that could just as easily have been stated by luminaries of the modern era. Not that science hasn't given astrology the benefit of the doubt; there have been numerous attempts to gauge the quality of astrological predictions. In each instance, without exception, predictions of astrologers have been no more accurate than chance alone would predict.

Despite a complete paucity of evidence, however, belief in astrology remains steadfast and unshakeable. Centuries after the Enlightenment, newspapers the world over still carry astrology columns awash with the kind of Barnum statements Forer cannibalized to make his famous test. It's no coincidence that these statements tend to be relatively flattering or empowering to the reader. Belief in astrology remains strong too. In 2010, 45 percent of Americans agreed with the statement that astrology was "sort of" or "very" scientific. Amusingly enough, it's been demonstrated that devout believers will still vouch that their reading is highly accurate even when shown the wrong reading.

As a mildly terrifying aside, there is also a thriving market for financial astrology, which applies this pseudoscience to financial markets for investors. Banker John Pierpont Morgan kept a personal astrologer on staff, allegedly stating that "Millionaires don't use astrologers, but billionaires do." This jaw-dropping practice continues to the present day; the chief technical analyst for HSBC stated in 2000 that "most astrology stuff doesn't check out, but some of it does," and another European bank stated that the correlation of astrological events and financial ones was "uncanny." This is rather frightening given the massive impacts of banking crises in recent years.

To reiterate just how useless financial astrology truly is, the British Association for the Advancement of Science fielded an experiment in 2001 where they gave a financial astrologer, an investor, and a five-year-old child £5,000 to invest on the FTSE100. Rather tellingly, the financial astrologer suffered the heaviest losses. Perhaps more concerning was the fact the child, choosing completely at random, outperformed the other two. Leaving aside the unlikely possibility of a savant at work, this demonstration is a rather damning indictment of the ability of financial astrologers, and arguably investors too.

Earlier, we considered psychics' adept use of reassuring platitudes, with their often employing the "rainbow ruse" technique, where they make a statement that covers an entire spectrum of contradictory personality traits, such as: "You are kind and compassionate, but if you feel betrayed your anger and resentment can be overpowering." Despite the intrinsic vapidity of such sentiments, evidence suggests many subjects identify more strongly with hollow meaningless appraisal rather than one specifically tailored to them. By utilizing statements such as these, the medium can steer the reactions of the subject and read them, projecting the impression that they have access to some intangible arcane knowledge when in fact they know nothing of the subject at all. The combined effect of these tricks can be extremely convincing—modern mentalists like Derren Brown and Ian Rowland use techniques to generate seemingly uncanny insight into people's lives, despite being completely upfront about their absolute lack of mystical ability.

Barnum statements aren't solely the preserve of the paranormal. Like Forer's original demonstration, personality tests provide a fertile ground for this effect. Perhaps the most pervasive example is the Myers–Briggs Type Indicator (MBTI) test beloved by many institutions. This purports to measure psychological styles in how people make decisions and is used to measure aptitude for jobs, personal training, and even marriage counseling. But despite the enthusiasm with which MBTI has long been deployed, studies to date indicate that it has very poor validity and fails to measure what it claims to. Were this not enough, it has low reproducibility, and the same subject can yield radically different personality results even a few days apart. Much of the criticism of these tests has focused on the vague nature of the testing questions, which lend themselves to the Forer bias of assigning a high rating to positive descriptions. The defects of this test prompted psychometric specialist Robert Hogan to opine that "most personality psychologists regard the MBTI as little more than an elaborate Chinese fortune cookie."

The Forer effect explains why we gravitate toward anything that indicates personal meaning in random noise and why we tend to fit generic statements to ourselves. But a more curious trait of human psychology is the revelation that expectations and beliefs alone can shape our perception of reality. Nothing demonstrates this odd quirk better than the *placebo effect*, where a patient reports tangible improvement in response to a sham treatment. Pain response is an archetypal example of this. Consider the 1996 experiment where participants had the index finger on one hand coated with the topical anesthetic trivaricaine while the other index was left untreated. Both were squeezed in a vice, and subjects assessed the pain levels. Predictably, the treated finger invariably hurt less. Yet trivaricaine was a fraudulent concoction, consisting only of water, iodine, and oil. To greater or lesser extents, we are all subject to the placebo effect, and the expectation of improvement is often enough to trigger some level of it.

This effect can even alter some physiological markers despite consisting of no active ingredient. Other research has shown that the more extreme the fake intervention is, the more we perceive it to be efficient, with sham injections and surgery having an even greater impact on perception than sugar pills. In this regard, the placebo is a self-fulfilling prophecy and a testament to the power of expectation. However, it is important to realize that placebo benefits are entirely perceptual; it is not a mystical example of mind over matter, nor can one cure conditions by wishful thinking. A perception of improvement might help with a cold or general aches and pains, but it cannot ever circumvent the need for medical interventions in more serious conditions. To this day, placebo usage in medicine is still a hotly contested ethical debate. As we'll see in later chapters, much of the placebo effect's apparent power can be explained by basic statistics.

The word "placebo" itself derives from a translation of the Bible by St Jerome, roughly equating to "I shall please." In medieval France, it was a funeral custom for a bereaved family to distribute largesse to mourners. This attracted obscure relations and outright pretenders feigning incredible anguish in the hope of monetary reward or food at the very least. Parasitic displays became widespread, with those simulating grief chanting St Jerome's *placebo Domino in regione vivorum*—leading to them to be disparagingly referred to as "placebo singers." The term soon traversed the Channel; the character of Placebo in Chaucer's *Canterbury Tales* is an unreformed sycophant, embodying these loathsome traits. In the era of pre-scientific medicine, placebo interventions were common in lieu of effective cures; the pioneering fifteenth-century French barber surgeon Ambroise Paré commented that a physician's duty was: *Guérir quelquefois, soulager souvent, consoler toujours* ("cure occasionally, relieve often, console always").*

The placebo effect explains some of the enduring popularity of a plethora of bunk or inert treatments, from reiki to iridology to craniosacral therapy and beyond. These treatments have no physiological basis but provide an illusion of relief in some instances.

---

* Despite living in pre-scientific times, Paré employed a version of the scientific method to evaluate claims, practicing evidence-based medicine. He invented surgical implements and greatly advanced battlefield medicine, at the time frequently beyond the skill of medics. This often led to soldiers' euthanizing their profoundly wounded colleagues; at the battle of Milan in 1536, he encountered two men horrifically burned by gunpowder. After admitting he could do nothing for them, one of their fellow soldiers unsheathed his dagger and slit their throats. The horrified Paré berated the soldier, who calmly replied that he would wish for the same in their position.

It may therefore sound like harmless quackery—after all, people have been buying snake oil from devious charlatans or elixirs from well-meaning but misguided folk for centuries, surely rendering it a mere harmless indulgence. Alas, alternative practitioners frequently shun conventional medicine, even informing their clients that such treatments are dangerous. And as they are usually unqualified to practice medicine, they all too frequently miss the glaring warning signs of serious illness while denigrating conventional medical intervention. This has led to patient deaths, most tragically of children whose parents were advocates of alternative medicine. Examples could fill numerous books—www.whatstheharm.net alone lists thousands of cases where patients have died or suffered at the hands of alternative practitioners.

To compound this, many in the alternative-medicine community are profoundly against life-saving interventions like vaccination; one study of homeopaths found that 83 percent urged their clients not to get immunized, instead selling them clinically useless concoctions for killer diseases such as malaria and measles. This is deeply dangerous, totally unregulated, and utterly misguided. Understandably perhaps, people are afraid of the side effects of pharmacological interventions and practices, and this is perfectly valid and often well founded. It does not, however, boost alternative medicine to anything other than an elaborate placebo, entirely incapable of treating any serious condition. Alternative medicine, then, is at best useless and at worst positively damaging, not just medically but to our collective understanding of science. By encouraging followers to cling to these delusions, advocates of alternative medicine denigrate the enormous strides we as a species have made over the past century or so in understanding the world around us and how our bodies work.

The modern medical usage of placebo arrived with the dawn of scientific medicine. The term was applied to medicine in 1920 in *The Lancet* by T. C. Graves, who wrote of ostensibly useless interventions where "a real psychotherapeutic effect appears to have been produced." Researchers quickly realized that an excellent way of testing new drugs was to randomly sort trial participants into groups, giving one cohort the drug to be tested and another the placebo. When the results are analyzed, they can then be used to ascertain if a given compound has an effect beyond mere suggestion, and indeed this is in effect how a double-blind (meaning neither researchers nor subjects know who is getting which agent) placebo-controlled trial is conducted—a gold-standard method of assessing the efficacy of new medications.

The placebo effect may underpin some perceived benefit from a host of inactive treatments. But can the effect occur in reverse? Could one be convinced that a given intervention is harmful, even if it is completely benign? The answer is yes. If one is sufficiently swayed to believe that something is detrimental, then by the same psychological mechanism one is inclined to display negative reactions to the inert agent. This inverted cousin of placebo is known as the *nocebo effect*, and it is arguably even more potent. Like its benign cousin, it too shares a Latin root: "I shall harm." While the term itself was only coined in the 1960s, the concept had been used at least as far back as the 1500s, when church authorities would give those suffering from alleged demonic possession fake holy relics; if the afflicted reacted violently to these frauds, their possession was deemed to be a folly of the mind rather than anything supernatural.

A contemporary example of this phenomenon manifests in electromagnetic hypersensitivity (EHS). Sufferers of this condition report an allergy to electricity or electromagnetic radiation (EMR), with a host of curiously varied symptoms, including fatigue, sleep disturbance, generic pains, and skin conditions. This is extremely debilitating, for there are few things as ubiquitous as electromagnetic radiation—from the familiar visible light that illuminates all we see, to the broadcast media transmitted across the globe by radio. For those afflicted with EHS, it is typically modern communications that yield the most reported malaise. Belief in EHS is strong and sincere, with an array of dedicated support groups around the world. There is no shortage of dubious health gurus making noxious claims about the dangers of radio frequency radiation either, and we encountered misguided cancer links with 5G and Wi-Fi in Chapter 2. While cancer is a very stark assertion, claims of electrosensitivity tend to be more nebulous, and those who promote such beliefs frequently hustle snake-oil cures to the ostensible illness.

Such is the depth of feeling on the issue that sufferers have even mounted high-profile legal action; in Santa Fe, activist groups tried to get public Wi-Fi hotspots banned on health grounds. A 2014 case saw a Massachusetts family filing a lawsuit against their son's school, contending that Wi-Fi there was making him ill. In a particularly tragic case in 2015, parents of 15-year-old Jenny Fry claimed that EHS had driven her to suicide, launching a campaign to remove Wi-Fi from UK schools. The same year, a French court ruled that an EHS sufferer was entitled to disability benefits. Sufferers even relocate to take refuge from their affliction. Towns with mandated radio quiet zones for astronomical research and military reasons, like Green Banks in Virginia, are besieged by EHS sufferers in search of respite, often leading to conflict with locals.

There is no doubt that sufferers endure real distress but, despite their assertions that EMR is the cause of their woes, there is plenty of evidence that the illness is wholly psychosomatic. Perhaps the strongest evidence lies in provocation studies, where those with hypersensitivity are exposed to varying sources of EMR to provoke a reaction and gauge the response. In trials to date, sufferers have been entirely unable to distinguish between real and sham sources. Their reactions are consistent only with belief, with sham sources, possessing no viable EMR, triggering a reaction. Similarly, sufferers do not report symptoms where they are unaware they are being exposed to a real source of EMR. This result has been replicated in numerous trials, and the inescapable reality is that EHS has nothing to do with EMR, and everything to do with our curious psychology. The WHO report on EHS, while sympathetic, is unequivocally clear: "The symptoms are certainly real and can vary widely in their severity. Whatever its cause, EHS can be a disabling problem for the affected individual. EHS has no clear diagnostic criteria and there is no scientific basis to link EHS symptoms to EMR exposure."

From a physics perspective, even if EHS were not an artifact of nocebo, microwave photons would make a strange suspect. As we've already seen, they are thousands of times less energetic than photons of visible light. The fixation on microwaves in EHS would seem odd, as the genesis for these fears is an unlikely one: the humble microwave oven. Microwave ovens are adept at heating food through a process known as dielectric heating. Molecules like water have regions of partial positive and negative charge, which in the presence of an electric field rotate to align themselves in the same direction. A typical domestic microwave oven emits photons with a frequency of approximately 2.45 GHz. This means that these particles of light flip electric field polarity 2.45 billion times a second, causing polar water molecules to rapidly bump off each other as they attempt to align themselves to the rapidly oscillating field. These rapid collisions yield friction, and consequently heat our food. This mechanism explains why microwaves are so efficient at cooking predominantly water-based food and conversely terrible at heating dehydrated substances.

This useful quirk of microwave energy is unfortunately ripe for misunderstanding, and a plethora of dubious gurus assert that microwave-cooked food is harmful by dint of being "exposed" to radiation. But this is nonsensical: Microwaves are not radioactive. They do not "irradiate" food but instead harness vibrational energy to heat it. Misguided extrapolation yields needless concerns: If microwave ovens can cook meat, then are our Wi-Fi routers and mobile phones frying us? This fear is again grounded in misunderstanding; the power output of our communication technology is thousands of times below that of ovens, with typical home routers outputting less than 100 milliwatts. Indeed, ovens are specifically designed to concentrate high-power microwave radiation using specially designed waveguides, magnetrons, and reflective chambers. This situation is neither encountered nor desirable in our conventional communication technology.*

Despite these facts, the sheer volume of odious claims about the dangers of Wi-Fi and mobile phones skews our risk perception. To compound this, while we are familiar with modern technology, we lack an appreciation of how it really works. When these factors are coupled together, it is hardly surprising that microwave radiation is such a magnet for the nocebo effect. That EHS is psychosomatic rather than physiological in origin does not make it feel any less real to the afflicted, even if they are mistaken about the cause of their woes. Sadly, sufferers reject the overwhelming evidence that their condition is psychological. Instead of trusting in good science, they cling to the statements of self-proclaimed authorities, with scientific evidence frequently dismissed as the product of conspiracy or ineptitude. For example, Sarah Dacre, the head of ElectroSensitivity UK, stated: "Conventional government-funded science isn't a reliable indicator of health defects. There's a vested interest in keeping the truth out of circulation."

As we've seen already, this adage is depressingly common. Claims of conspiracy are a reassuring fall-back into confirmation bias in lieu of reevaluation. The nocebo effect underpins a wide variety of other similar illnesses. The perennial onslaught of dubious claims about water fluoridation serves as a resonant example. Despite decades of safety data showing fluoride to be a safe and effective addition to

---

* In any case, the intensity of a spherically emitted electromagnetic radiation source has an inverse square relationship with distance; at a distance of two meters from a source, the intensity is only one-quarter that of a distance of one meter. At three meters, field intensity is only one-ninth the magnitude. This physical law means the strength of an EMR source diminishes enormously over even modest distances, even for naked Wi-Fi sources.

water to improve dental health, there is a vocal worldwide network adamant that fluoride is responsible for all manner of malady. Here too, there's evidence of the nocebo effect; in Finland in 1992, aggressive protests against water fluoridation led the Kuopio city council to remove it from the water supply. However, by way of experiment, the fluoride was not removed from the water supply on the date announced but instead at a different time. Surveys taken indicated that people reported ill effects from water only if they thought it contained fluoride, irrespective of whether it actually did—further proof of the self-deluding power of expectation.

This is a familiar story, of course; the anti-vaccine movement focuses its narrative on anecdotes of harm post-inoculation, where the nocebo effect frequently rears its ugly head. You might reasonably ask how people remain so convinced of dubious positions when expert opinion is so against them. Conspiratorial narratives certainly form part of this, but cognitive illusion too must play a role. In a wonderfully titled 1999 paper, "Unskilled and Unaware of It: How Difficulties in Recognizing One's Own Incompetence Lead to Inflated Self-Assessments," psychologists David Dunning and Justin Kruger observed how those with low ability or expertise in a given subject mistakenly assess their cognitive ability or knowledge as vastly greater than it actually is.

The Dunning–Kruger effect most certainly manifests in the anti-vaccine community; a 2017 paper asked subjects to rate their knowledge of what causes autism, contrasting this to medical and scientific professionals' knowledge. It further asked to what extent they agreed there was a link between vaccination and autism. The results made for depressing reading: 62 percent of those who performed worst on the autism knowledge test believed they knew more than the medical community, and 71 percent of those strongly endorsing the link between vaccines and autism also asserted superior knowledge on the subject.

Such results are known in social psychology as examples of illusory superiority, the overestimation of one's ability in relation to others. They call to mind Bertrand Russell's dictum that "the fundamental cause of the trouble is that in the modern world the stupid are cocksure while the intelligent are full of doubt." This is perhaps a substantial reason why absolutists and fundamentalists of every ilk hold disproportionate sway over public perception. The tragedy of modern discourse is that all too frequently, one's confidence to opine on a subject is inversely proportional to one's understanding of the topic at hand. Nor is this lack of expertise any impediment to strong opinion or forceful assertion. And to our detriment, misguided confidence more frequently begets public attention than uncertain knowledge. In reality, total objectivity is an ideal that we seldom reach. Our expectations invariably shape perceptions and reactions. Our affinity for fortune-tellers and horoscopes is a product of our need for validation, and the mere suggestion of an effect—whether positive or ill—is often in itself enough to prompt a visceral reaction.

But even if we're not consciously aware of it, we are deeply social creatures, and this influences us more than we might acknowledge. Our expectations, which so powerfully shape our perception, are themselves shaped not only by the people around us, but by what we are exposed to. In particular, the media, advertising, and information we consume has a huge effect on us—it is no coincidence that fears over vaccination spike with media coverage, or that EHS and fluoride fears are fueled by an underbelly of pseudoscience blogs.

Of course, we live in the age of data, and numbers have a tangible effect on our perceptions. We're bombarded daily by numbers, statistics, and trends from which we are supposed to make sense of the world. While this is vital to our well-being, we are collectively rather numerophobic, and sometimes even seemingly obvious trends hide trapdoors that confound us. Just how important is this influence, and how can it drive us toward dangerously error-strewn conclusions? That is a complex but important question—and one we will tackle in the next few chapters.

PART 4

Lies, Damned Lies, and Statistics

How Numbers Can Mislead Us

"Politicians use statistics in the same way that a
drunk uses lampposts—for support rather than
illumination."

—ANDREW LANG

12

CHANCE ENCOUNTERS

Bayes's Theorem, Gambling, Wrongful Convictions, and
How Misunderstanding Probability Can Lead to Tragedy

The aptly named author and playwright Marilyn vos Savant is famous for her intellect*; between 1986 and 1989, vos Savant held the Guinness world record for highest recorded IQ. Guinness eventually retired the category when it became clear the psycho-metric tests used were completely unreliable, but vos Savant's high intelligence was never in doubt. Her fame led to a weekly column for *Parade* magazine, where readers would invite her to answer questions of logic and solve puzzles. In 1990, Craig Whitaker of Maryland posed the following question:

> Suppose you're on a game show, and you're given the choice of three doors. Behind one door is a car, behind the others, goats. You pick a door, say #1, and the host, who knows what's behind the doors, opens another door, say #3, which has a goat. He says to you, "Do you want to pick door #2?" Is it to your advantage to switch your choice of doors?

This strange question was loosely based on a dilemma faced by con-testants on the game show *Let's Make a Deal*, where host Monty Hall would offer a choice to switch or stay. To most people, the answer seemed rather obvious. If there are two doors remaining, then surely it's 50/50 regardless, and whether one switches or stays is irrelevant? This was not the answer vos Savant gave. Instead, she advised that switching was the most advantageous tactic. A deluge of furious letters followed, denigrating her ignorance. Of the more than 10,000 missives received on the subject, roughly 1,000 came from individuals

---

* Apropos of nothing, vos Savant is married to Robert Jarvik, whom we encountered before.

with doctorates, many from mathematicians and scientists. These letters, steeped in condescension, castigated her for perpetuating public innumeracy.

But vos Savant was correct: switching doors after the host's reveal would give the contestant a two-in-three chance of winning, relative to an only one-in-three chance of victory if one stays with the initial choice. Had those condemning vos Savant's ignorance cared to look deeper, they might have seen that this "Monty Hall Problem" had been posed and solved by statistician Steve Selvin in 1975. How can this strange result be true? Pretend the car was behind door A. If you were to pick that door, Monty would reveal the goat behind either door B or C. If you switched in this instance, you'd lose. But imagine instead you picked door B first; Monty would open door C, and switching would win you the car. Similarly, if you picked C, door B would be unveiled, and again switching would be your winning strategy. Two-thirds of the time, switching is one's optimal strategy.

	PICK A	PICK B	PICK C
CAR IN A	*STAY WINS*	SWITCH WINS	SWITCH WINS
CAR IN B	SWITCH WINS	*STAY WINS*	SWITCH WINS
CAR IN C	SWITCH WINS	SWITCH WINS	*STAY WINS*

The pay-off matrix above lists every possible configuration and, in two-thirds of cases, switching rather than staying is the winning strategy. This seems patently absurd because intuitively we feel there shouldn't be a difference whether one switches or stays. If you found the result perplexing, you're in good company. Aside from the furious readers of *Parade* magazine, prolific mathematician Paul Erdős remained dubious about the veracity of the solution, eventually being won over by computer simulation. Today the Monty Hall problem is a cornerstone of probability textbooks and yet still bamboozles even experts. Curiously, experiments with pigeons indicate they rapidly learn that the switching strategy is optimal. This stands in stark contrast to humans; as the experimenters drily noted: "Replication of the procedure with human participants showed that humans failed to adopt optimal strategies, even with extensive training."

Our innate ability to seek out and quantify patterns in all we encounter is one of our finest survival skills. Our urge to make sense of the world around us and our insatiable curiosity have led us as a species to civilization, great discoveries, and virtual mastery of the physical world around us. Yet this fine instinct can wholly fail us when confronted with the noisy and chaotic patterns we face every day. In an uncertain world, probability and statistics wielded wisely act as a cleaving blade, separating the real from the illusory. Chance events can be understood as probability, an area of fundamental importance in everything from city planning to quantum mechanics, medical research to economics. Despite the high-minded applications of statistics and probability, the origin of these techniques can be traced back to a more earthly motivation: gambling. Humankind has enjoyed games of chance for millennia, but until the seventeenth century the foibles of the dice were considered far beyond the realm of man, arcane providence firmly seated in the lap of the gods. The idea that outcome could be predicted to some degree of accuracy seemed impossible, even vaguely blasphemous.

So it may have remained, had a curious problem posed by the eccentric French writer the Chevalier de Méré in 1664 not captured the attention of two of the finest minds of seventeenth-century France—Blaise Pascal and Pierre de Fermat. Pascal ultimately solved de Méré's problem, proving it is very slightly more probable to roll at least one six in four rolls of a single die (51.77 percent chance) than it is to role at least one pair of sixes in 24 rolls of 2 dice (49.14 percent chance). This difference seems negligible, but it is crucial—betting on the former case, a gambler would expect to profit over time, while they would ultimately run a loss betting on the latter instance. In the saloon culture of pre-revolutionary France, great minds poured their ample faculties into the evasive gambler's grail of maximizing profit.

Their investigations into dice games led to the birth of probability theory, emerging from humble parlor games. But as we've seen, our hair trigger is finely honed for pattern finding, and consequently we tend to have a flawed perception of the stochastic, or randomly determined. Truly random events have no "memory" of previous outcomes, yet our human tenacity to extrapolate from our observations often tempts us to erroneous conclusions.

Take, for example, a lottery. If things are fair, the numbers 1, 2, 3, 4, 5, and 6 are just as likely to tumble from the machine as any other combination. Still, this intuitively feels less likely than a wider spread of numbers, and most of us would avoid it. Similarly, if a fair coin is flipped 20 times and comes up heads every time, we expect the tails to be "due" on the 21st flip, even though the probability remains exactly 50 percent.* This is the *gambler's fallacy*, underpinning the ruin of many. Thankfully though, we are not completely at the mercy of instinct, useful yet often wrong-headed as it is; human curiosity over the centuries has led us to develop tools to distinguish noise from signal.

In the twenty-first century, statistical and probabilistic information is ubiquitous, conveying information about everything from markets to medicine, sports results to weather patterns. Statistics hold an appeal in part due to their seeming intuitiveness. Yet this veneer of simplicity is often misleading, hiding subtleties that can completely derail us. The opacity of statistics and widespread innumeracy makes it all too easy for statistical trends to be misinterpreted by the unwary. More alarmingly, this same ambiguity also allows us to be manipulated by schemers to bolster fallacious arguments. All of this is to our collective detriment, and cynicism about statistics is easy to sympathize with; see the famous quip about there being three kinds of lies: lies, damned lies, and statistics. First attributed posthumously to Prime Minister Benjamin Disraeli in 1892, it was popularized by Mark Twain in his autobiography.

* Although the probability of 20 heads in a row is 1/1,048,576, so we might reasonably start asking questions about how fair the coin is at this juncture.

Although cynicism is understandable, however, dismissing
statistics as mere Trojan horses for falsehood would be to
completely jettison the baby with the bathwater; statistician
Frederick Mosteller noted that "while it is easy to lie with statistics, it is even easier to lie without them." This is certainly
true—when applied correctly, statistical tools are invaluable,
unlocking secret trends that evade even our well-honed eyes.
This discerning power has rendered them invaluable in every
avenue from medicine to politics. Yet, if we are to benefit from
them, we need be wary of all the pitfalls we can stumble into
when presented with statistical information. Abuses occur all too
frequently when numeric information is invoked in argument.
We must refine our understanding so that we may circumvent
ineptitude or trickery.

At their best, statistics are incredibly useful at quantifying life
in an uncertain world. But at worst, devoid of context and understanding, they can be mystifying and misleading. To illustrate the
curious nature of statistics and probability, let's take a counterintuitive example illustrating both aspects.

Imagine you're given an HIV test, which you're informed is 99.99
percent accurate. The test comes back positive; then what are the
odds you have HIV? For most of us, our instinct quite reasonably
tells us it is almost certain we have the disease, yet this is generally
wrong. The actual answer is instead closer to 50 percent for most
of us. If you're left somewhat perplexed by that result, you'd be in
good company; most people, including medical professionals, tend
to be equally flummoxed by this seemingly bizarre assertion.

This curious result is a consequence of Bayes's theorem, a
mathematical framework for combining conditional probabilities, mapping how probability branches out. Bayes's theorem tells
us that the probability of having HIV in the event of a positive
test is dependent not only on the test but on how likely one truly
is to have the illness. While the test itself is almost perfect, its
accuracy is dependent upon another condition, namely the a

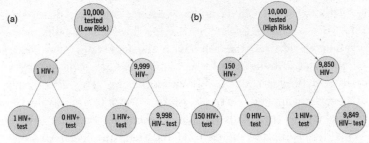

Frequency trees depicting reliability of HIV tests for (a) low-risk cohort and (b) high-risk cohort

priori chance that a person has the virus in the first place. We'll avoid a formal statement of Bayes's theorem as it is beyond our scope and is needlessly intimidating to those unfamiliar with mathematical notation. However, the logic behind it is easy to follow and vital to illustrate, as it lurks behind countless seemingly paradoxical statistics.

Returning to our example, how exactly can a test with 99.99 percent accuracy only be half sure a typical patient has HIV? For a typical low-risk subject, baseline infection rate is about 1 in 10,000. Now, imagine 10,000 such people walk in for a HIV test; one of them has the virus and will almost certainly test positive. But in the remaining 9,999, another will test positive due to the accuracy limits of the test, leaving two positive tests, only one of which is a true positive, meaning that with a positive test, a person is 50 percent likely to have the illness.

Crucially, this jarring result does not indicate that the test is inadequate; the HIV test in our example is incredibly accurate. Rather, due to the limited prevalence of the illness, the *conditional probability* is much lower than what we may intuitively expect. In truth, the a priori likelihood of a particular subject being infected is inextricably entangled with the precision of the result. Consider the same test administered to a high-risk population, such as intravenous drug users. The infection rate in this cohort is roughly 1.5 percent. Let's again envision 10,000 such patients getting tested. In this cohort, roughly 150 will have the virus and flag positive. Of the remaining 9,850 patients, there should be approximately one false positive. In this instance, the odds of HIV infection given a positive test are not 50/50 any more. The likelihood of a high-risk patient having HIV given a positive test is 150/151 or 99.34 percent—much greater than a patient in the low-risk cohort.

The low- and high-risk scenarios can be illustrated more intuitively with a frequency tree, depicted in the figure above. This difference is extreme, and it's worth dwelling on this finding for a moment. We might reasonably ask: Why the stratification? Why should the same test administered to one group yield an accuracy so drastically different from another group with the same test? Instinctively we may feel there is something wrong with the test, but this is not the case—the test does not discriminate, nor does its inherent precision selectively improve or disimprove given a patient's background. The needle is not clairvoyant, remaining 99.99 percent accurate for all patients. The crux of the issue is that Bayes's theorem shows us that this information on its own will never be enough to draw conclusions on issues that depend on other probabilities. Probabilities are often conditional, and naked numbers devoid of context need to be carefully parsed.

This serves as an illustration of the fact that, despite the ostensibly intuitive nature of probability and statistics, their seeming simplicity hides layers of complexity that are easy to misunderstand. Such misunderstandings can drive us to entirely erroneous conclusions, and dubious inference and statistical misunderstandings can all too frequently have detrimental consequences. The rationale behind these misunderstandings is no mere academic triviality, nor mathematical sleight of hand; we live in an age where statistical information decides policy in every arena imaginable, including science, politics, and economics. Yet the very ubiquity of statistics and probability in our lives means that they often decide matters of life and death, be they medical treatments or government action.

In these cases, our very lives might depend on drawing the right conclusion from probabilistic information. When mistakes are made, especially by those who should know better, there can be a high human cost. In the early days of the AIDS crisis, before the advent of antiretroviral drugs, a positive HIV status was considered tantamount to a death sentence. The sheer reliability of the HIV test lured many doctors into a false sense of confidence, and many patients were told it was practically certain they had the illness when they did not, leading many to spiral into depression and reckless behaviors over a false positive that was surprisingly likely to occur.

There is another arena where probability decides the fate of many a person: the courtroom. Juries and judges are given the unenviable task of ascertaining guilt. To reach their conclusions, they are frequently accosted with a barrage of statistical information from both prosecution and defense. In any adversarial legal case where statistics are invoked, both sides have a vested interest in presenting this information—namely, their client. Their information is presented to juries to sway them one way or another but, as we've seen from the HIV testing example, frequently these numbers in isolation tell us practically nothing and are liable to misunderstandings, even leading juries to a conclusion at odds with the reality of the situation. Statistics might make good sound bites, but divorced of qualifying information, they're as likely to mislead as to enlighten.

To see just how tragic the consequences of such incompetence can be, one need only consider the debacle of Professor Sir Roy Meadow. Meadow is a distinguished British pediatrician, famous for his 1977 academic paper on Munchausen syndrome by proxy. Knighted for his contributions to child health, his thinking on the subject was for a time hugely influential on social workers and the National Society for the Prevention of Cruelty to Children. He was famed for a dictum that became eponymously known as "Meadow's law": "One sudden infant death is a tragedy, two is suspicious, and three is murder until proved otherwise."

Yet Meadow's penchant for seeing dark forces at play everywhere was the result of statistical ineptitude, and his innumeracy destroyed lives. Nowhere is this more obvious than in the atrocious ordeal suffered by Sally Clark in the late 1990s. Sally and her husband Steve, both lawyers, had endured the devastating misfortune of losing two young sons to what appeared to be sudden infant death syndrome (SIDS). Their first son, Christopher, had fallen unconscious and died at the age of only 11 weeks. Their second son, Harry, passed away under similar circumstances at the age of eight weeks. In both instances Sally had been alone with the children. There had been some arguable signs of trauma, potentially due to her frantic attempts to resuscitate the boys. This was enough to place her under suspicion.

Exacerbating the incredible grief the Clarks faced, both Sally and Steve were charged with murder. With scant physical evidence, the case against Steve was dropped, but the Crown opted to continue with Sally's trial. Meadow—at that time considered the UK's foremost authority on child abuse—was brought in by the prosecution to give evidence against her. Due to the paucity of physical evidence, Meadows made a statistical argument for guilt; he asserted that, for a middle-class nonsmoking family like the Clarks, the likelihood of an occurrence of SIDS was 1 in 8,543. Thus, he reasoned, the chances of two cases of SIDS in the one family was , roughly 1 in 73 million. Meadows likened this to the jury as akin to an extremely outlandish race win:

> It's the chance of backing that long-odds outsider at the Grand National, you know; let's say it's an 80 to 1 chance, you back the winner last year, then the next year there's another horse at 80 to 1

and it is still 80 to 1 and you back it again and it wins. Now here we're in a situation that, you know, to get to these odds of 73 million you've got to back that 1 in 80 chance four years running, so yes, you might be very, very lucky because each time it's just been a 1 in 80 chance and, you know, you've happened to have won it, but the chance of it happening four years running we all know is extraordinarily unlikely. So it's the same with these deaths. You have to say two unlikely events have happened and together it's very, very, very unlikely.

The figure was assured, stark, and unambiguous—a seeming smoking gun placed in Sally Clark's hand. Unsurprisingly, the media snapped this sound bite up as unassailable proof of guilt and so too did the jury. Based largely on Meadow's testimony, Sally Clark became a figure of hate, vilified by the popular press as a remorseless child-killer. The jury reflected this public mood, ultimately convicting her of double infanticide.

Yet the verdict horrified statisticians, for good reason—to arrive at the figure of 1 in 73 million, Meadow simply multiplied the probability of two independent events together. This is perfectly correct when dealing with events like coin flips and roulette wheels, where each outcome is truly independent of the previous one. But it fails horribly when the events are not independent. Even in the late 1990s it was well known from epidemiological data that SIDS tends to run in families, perhaps due to genetic or environmental factors. This renders the blithe assumption that the two deaths are independent absolutely nonsensical, and the probability brandished against Clark was dubious in the extreme.

Nor was this the only error at play. Compounding Meadow's woefully inept usage of statistics, both the jury and national media's assumption of Sally's guilt rested on a statistical faux pas so common in courts that it is known as the *prosecutor's fallacy*. Let's pretend for a moment that Meadow's figure had been correct. This would be interpreted by many as equivalent to the probability of her innocence being 1 in 73 million. Yet this inference is completely off base. While multiple cases of SIDS may be rare, so too are multiple maternal infanticides. To work out which is more likely, these competing explanations need to be compared in order to determine their relative likelihood. In Clark's case, if such an analysis had been properly performed, the likelihood of two SIDS tragedies would have been greater than the murder hypothesis, illuminating the inherent problem with the prosecutor's fallacy.

The grave injustice against Clark did not go entirely unnoticed. In an exceptionally strongly worded and comprehensive rebuke, the Royal Statistical Society (RSS) slammed the prosecution's abuse of statistics, pleading with the Lord Chancellor to consider the case more carefully. Stephen J. Watkins, then editor of the *British Medical Journal*, wrote a damning editorial on the misuse of medical statistics in the case, opining that "defendants deserve the same protection as patients." Sadly, their protestations fell on deaf ears. It is impossible to comprehend what Sally Clark must have gone through. The grief of losing her children coupled with the rank injustice of her false conviction must have been devastating. She was demonized in the press, condemned from pulpits. In prison, her nightmare was worsened by other inmates ostracizing her for the nature of her ostensible crime, and also because she had been a lawyer and the daughter of a police officer.

Worryingly, this appalling miscarriage of justice might never have been rectified had it not been for the dedicated efforts of a handful of people. Steve Clark left his partnership in Manchester, taking a position as a legal assistant near Sally's prison, selling the house to finance legal fees and appeals. He was joined in his efforts by renowned lawyer Marilyn Stowe, who volunteered her services pro bono, as her professional judgment convinced her that the case against Clark was nonexistent. Sally's vindication was an arduous process, with the first appeal judges downplaying the demonstrations of statistical incompetence as little more than a mathematical trick.

In the end, it was Stowe's dogged and resourceful nature that forced the second appeals court to pay heed. Through some impressive sleuthing, she determined that microbacterial examinations by the prosecution's pathologist, Alan Williams, had uncovered evidence of a *staphylococcus aureus* colony in Harry's autopsy. This strongly suggested that it was a contributory factor to his death, which had not been made known to the defense or investigating officers. Due to the efforts of Marilyn Stowe and Steve Clark, Sally's conviction was finally overturned in 2003. The second appeals court conceded that the statistical errors had utterly perplexed the jury and hopelessly biased the trial: "We rather suspect that with the graphic reference by Professor Meadow to the chances of backing long-odds winners of the Grand National year after year it may have had a major effect on [the jury's] thinking."

The exoneration of Sally Clark had a domino effect. Meadow's sterling reputation as a virtually infallible expert witness was severely undermined, prompting a slew of reviews of cases in which he had given evidence. This led to a number of other women imprisoned as a result of his dreadful statistics being released from jail. By the time she was released, Sally Clark had spent more than three hellish years in jail and was profoundly affected by the ordeal. Steve sadly noted that she would "never be well again," suffering from protracted grief and a number of serious psychological disorders. Sally Clark died in 2007 of acute alcohol intoxication, her life irreparably damaged by the failure of both experts and the general public to comprehend statistics.

The tragic story of Sally Clark is a potent reminder that numbers matter. It is of paramount importance to understand that statistics divorced of context and qualification are fertile grounds for confusion. It is deeply unsettling to think of the legal cases that have hinged on dubious statistical inference, or the number of innocent people convicted on mathematical misunderstanding. Even excellent science can lead to atrocious inference; consider DNA profiling and its unparalleled ability to read the very code from which we are written.

The sheer power of DNA evidence means it is often presumed by public and legal experts alike to be beyond reproach, incapable of error. Yet, while it is undeniably true that DNA evidence is a powerful tool in bringing offenders to justice, it is not infallible and is as subject to errors as any other scientific investigation.

Like our HIV-test example, the reliability of conclusions drawn from DNA evidence depends on a priori information about the case at hand. For example, pretend a partial DNA profile discovered at a crime scene occurs with a frequency of roughly one in a million. If we have a suspect in custody who matches this profile, this may be considered strong evidence against them, or a "hot hit." However, if we instead were to trawl through a huge database of 10 million people, we'd expect to find ten "cold hits," simply due to coincidence. Bayes's theorem demands that, in order to gauge probability of guilt, we would need to consider not only the test results but the frequency and sample size from which it derives. Thus, the strength of a given piece of DNA evidence is dependent upon whether it was obtained from a single subject or from a database hit. Without that information, juries risk engaging in the prosecutor's fallacy.

Again, this is not a limitation of the technology but rather of how we interpret findings. It is undeniable that DNA evidence has revolutionized legal proceedings, yet careless interpretation can lead—and has led—to false incrimination, so great care has to be taken to avoid these probabilistic pitfalls. While probability might seem superficially straightforward, the reality is that the appearance of intuitive simplicity is often completely illusory. To truly understand what numbers tell us requires context and consideration, and sometimes the true message they convey might be completely at odds with our initial impressions.

The paradoxical nature of statistics means that seemingly obvious trends may misguide us, even when the data seems to support a particular hypothesis. We intuitively believe that the numbers speak for themselves, but we often forget that they require some interpretation.

13

SIFTING THE SIGNAL

*The Theranos Controversy, Cholera Outbreaks, Why Correlation
Does Not Equal Causation, and Other Statistical Nightmares*

In 1973, the University of California, Berkeley, was sued for sex discrimination. The evidence, on the face of it, seemed pretty damning. Of males who applied to the esteemed UC campus, 44 percent were accepted, while only 35 percent of women applying gained a place. Such a disparity appeared rather suspicious, indicative of underlying sexism in the admissions process. Accordingly, a legal challenge was mounted to expose and counteract this bias. But the ensuing investigation unearthed a curious result: When the admissions data was analyzed, there was a "small but statistically significant bias in favor of women across most departments."

How can this be the case when these two positions seem mutually contradictory? If women were as likely (or even slightly more likely) to be accepted into a given department as males, why then is this not reflected in the initial statistic? The solution to the paradox reveals itself when one looks a little more deeply into the stratified admission data. Hiding within it was a pattern not immediately apparent in the "percentage admitted" statistics. Males, on average, tended to apply for less competitive departments with high rates of admission among all qualified candidates, such as engineering. Female candidates, by contrast, tended toward departments like English, where competition was incredibly high even among well-qualified candidates.

The problem in Berkeley's case was not naked gender discrimination at admission, but a *lurking variable* (or *confounding variable*) of gender-specific selection in field, skewing overall rejection rates. As the study authors noted: "Measuring bias is usually harder than assumed, and the evidence is sometimes contrary to expectation." None of this is to deny the noxious role of sexism, and the authors to their credit noted that "absence of a demonstrable bias in the graduate admissions system does not give grounds for concluding that there must be no bias anywhere else in the educational process or in its culmination in professional activity." Behind the odd result in Berkeley lies Simpson's paradox, a counterintuitive phenomenon that an apparently clear trend in groups of data can disappear or even reverse when these groups are brought together.

One of the curious problems we face now is that getting data has never been easier, but a naive interpretation of the available information and trends leaves us with impressions totally at odds with reality. Simpson's paradox raises its head often in political, social, and medical fields, and occurs when causal relationships are incorrectly taken from frequency data. For example, the proportion of people who die in hospitals is much greater than those who die in post offices, but it would be absolutely wrong (and hopefully glaringly obvious) to infer that a post office is a better place for medical treatment than a hospital. Drawing a causal relationship from statistical data can be notoriously difficult; a single confounding variable can guide the unwary to completely false conclusions. A classic example is drowning deaths, which tend to increase with ice cream sales. The statistical relationship between the two is quite robust, but it would be outlandish to assume that ice cream causes drowning. The lurking variable here is simply good weather, which leads to both higher ice cream sales and more accidents in water.

Of all the rabbit holes one may fall into, few avenues are as rife with error as causation fallacies. We previously encountered the post hoc, ergo propter hoc school of logical fallacies, but in the context of statistical data these fallacies are not always quite as obvious as their rhetorical cousins. Whereas it's generally quite easy to spot causation fallacies in argument and discussion, our collective innumeracy means they can fly somewhat under the radar when statistics are involved. Despite the public's fetishizing of statistical information and trends, it is often forgotten that it is surprisingly difficult to make a robust causal connection. There are usually so many confounding variables that it takes a very carefully controlled analysis to work out the underlying relationship, if indeed there is one. The mantra that "correlation does not imply causation" must not be forgotten.

To separate cause from effect properly requires meticulous investigation. Correlation might provide a clue to some connection, but Simpson's paradox and the existence of lurking variables demonstrate why this information by itself must be treated cautiously. Interpreted the wrong way, this can establish totally incorrect narratives in the minds of the unwary. The statistician David R. Appleton and colleagues give a lovely example concerning mortality statistics for women in the English village of Whickham in the early 1970s, and a follow-up study twenty years later. Naive reading of the results seemed to suggest that smoking is in some way beneficial, as nonsmokers had a 31 percent mortality rate versus only 24 percent for smokers. This alarming result disappears when Simpson's paradox is considered; when results were broken down by age group, it showed that smoking was detrimental across all age groups. The confounding influence was that the smokers surveyed in the initial study tended to be younger than the nonsmokers in the sample, where few of the women over 65 smoked. As many of the older women had died by the time of follow-up, naïve interpretation of the statistics alone gives a false impression. With examples like this, it is easy to see how the unscrupulous can distort the truth by careful manipulation of the statistics.

Spurious relationships exist everywhere and, unless their confounding influence can be ruled out, it is premature and often wrong to assert cause from mere correlation. Correlation on its own needs to be carefully analyzed to ascertain true cause, even in the absence of confounding variables; one might correctly correlate umbrella usage with rain, but it would be wrong to assume that umbrellas cause that rain. Spurious relationships can be exploited for comic effect. Tyler Vigen finds strong correlations in completely disparate data sets, such as between US cheese consumption versus deaths due to people becoming tangled in their bed sheets, or suicides by strangulation versus the number of lawyers in North Carolina. Bobby Henderson, founder of the satirical religion Church of the Flying Spaghetti Monster, decreed that full pirate regalia is religiously mandated, pointing out the statistically significant inverse relationship between number of pirates worldwide and average global temperature, inferring that pirates prevent global warming.[*]

I should pause at this juncture, for fear I've inadvertently given the impression that statistical correlations are meaningless. In fact, nothing could be further from the truth. Statistical correlation can be viewed as an important element of a detective story. Imagine that a series of crimes has been committed. Statistical correlation might show that a suspect was in the area at the time of each crime. This on its own is not proof of guilt, but it is an excellent start to deciding whether further investigations are warranted. Similarly, if there is no relationship between the suspect's movements and the crimes, then we might disregard them. The only caveat is that such statistical tools should be well applied so that confounding influences can be avoided. Returning to our detective analogy for a moment, in a spate of murders there will probably be a correlation between the killer's movements and those of the coroner but—unless there's some firm reason to believe that the coroner has been moonlighting as a serial killer—it would be unwise to charge them on this evidence alone.

Statistical information must be carefully parsed to avoid grasping at false conclusions. In the mid-nineteenth century, the miasma theory of disease—where illness was thought to be spread by foul air—still dominated medical thinking. This belief was expressed blithely by the social reformer Sir Edwin Chadwick, who stated that "all smell is disease." As we've seen with malaria, belief in miasma

---

[*] Despite this being a deliberately outlandish parody, countries like Somalia with piracy problems do generally have lower CO_2 output than wealthier nations. The underlying reason for this is more likely to do with poverty and lack of industry, but I will take any pretext to swan around like a sixteenth-century buccaneer.

was ubiquitous and bolstered by the observation that outbreaks seemed to be accompanied by malodorous outpourings. Chadwick was a liberal champion of the poor in London, and by 1842 he had correctly identified sanitation as a major health issue. Under his supervision, the Metropolitan Commission of Sewers began gradually upgrading sewage systems throughout London, closing more than 200,000 cesspools.

Curiously, despite the miasma theory of disease being incorrect, the sewage reforms strengthened belief in it for a period, due to a very significant but misleading relationship: Cholera outbreaks decreased where cesspools had been closed. This was taken as confirmation of the belief that foul air was the vector for this and other illnesses. At around the same time, similar beliefs as to the origin of disease and devastating outbreaks led in part to the restoration of Paris and the Parisian sewer network. These outbreaks helped justify Georges-Eugène Haussmann's rebuilding of Paris from a cramped and dark city to one with beautiful spacious boulevards, luscious gardens, and clever planning—the City of Light as we know it today.

But even at this time, there were some who saw problems with the miasma theory. The physician John Snow was one such man. By 1854, the London sewage system had not yet reached the Soho district of the city, and a rapid influx of people meant that living spaces were tight. Cesspools had grown beyond capacity. On August 31, 1854, a vicious cholera outbreak struck around Broad Street, Soho. Within three days, 127 people were dead. Panic set in, and over the subsequent week roughly three-quarters of the residents had fled. By mid-September 1854, the outbreak had killed 500 people, with a mortality rate of 12.8 percent.

Conventional wisdom at the time dictated that bad air was the root cause, but Snow was not convinced. He instead began a thorough investigation, assisted by the Reverend Henry Whitehead. By talking to survivors and tracking victims' movements, a pattern emerged linking all cases: a single pump on Broad Street.

This must have seemed curious to Snow; while he harbored serious doubts over the miasma theory of disease, it would be another seven years before Louis Pasteur pioneered germ theory. Consequently, there was something of a void in the nineteenth-century understanding of epidemics concerning how disease was spread. Nevertheless, by employing pioneering statistical analysis in addition to his carefully plotted map, the pump aroused Snow's suspicion.

Of course, there were confounding variables. Local monks seemed unaffected, as did those who lived in the local brewery. Snow enquired further, finding that the monks only drank the beer they made, and likewise the brewery fermented any water taken. The fermentation process killed cholera bacteria, explaining the apparent immunity of the monks and brewers to the sickness.* There were some other strange outliers, such as a spate of deaths that occurred closer to a different pump. However, meticulous questioning by Snow and Whitehead revealed that in these clusters, the victims had deliberately taken water from the Broad Street well, preferring the taste. Taken together, these findings strongly suggested the that pump was the true cause. Presented with these findings, local authorities removed the pump handle, bringing the outbreak to a close.

In total, 616 people died due to the outbreak, but the rapid detective work of Snow and Whitehead undoubtedly saved more lives. More significant, perhaps, was the impact on science: The "ghost map" was a pivotal moment in epidemiology, the branch of science and medicine focusing on incidence, distribution, and causes of disease. It was a demonstration that even seemingly obvious correlations need to be tested, lest the wrong culprit be framed. From a medical science perspective, the Soho cholera outbreak was also the death knell for miasma, as Snow's pump aptly demonstrated that water could vector disease, in stark opposition to the mantra of air as the only possible source of poison. The discovery

---

* Important observation regarding water quality: When in doubt, opt for beer.

of microorganisms just a few years later was the final nail in the coffin for the obsolete theory, and a gateway into modern medicine.

The cause for the outbreak was much later determined to be due to the unpleasant reality that the Broad Street well had been placed just over 3 feet (1 m) from a cesspit, where infected fecal bacteria had seeped into the water supply, propagating violently. There is a curious footnote to all this that will perhaps be familiar to astute observers of politics. Once the immediate danger had subsided, the local authorities vehemently rejected Snow's evidence and replaced the pump handle, despite the very real danger of a fresh outbreak. This sordid rejection was born out of pure squeamishness and political considerations, as accepting the virtually incontrovertible evidence meant accepting the possibility of fecal-oral transmission. This was felt by the local administration to be too distasteful for the public to comprehend, demonstrating the long-standing and depressingly consistent habit of politicians through the ages to care more about public opinion than good evidence, often to the detriment of that very public.

Unwillingness to accept unpleasant statistical data isn't solely the preserve of politicians, however, and its occurrence certainly hasn't diminished with time. The glorious rise—and ignominious fall—of Silicon Valley darling Elizabeth Holmes is a much more contemporary tale. As a youth, Holmes was precocious, demonstrating an early entrepreneurial streak, and by high school she had started her first business, selling C++ compilers to Chinese universities. In 2004, at the age of 19, she dropped out of Stanford, using her tuition money as seed-funding for a new venture. This start-up had a lofty goal: to revolutionize health care. To capture the scope of her ambition, Holmes chose a portmanteau derived from the words "therapy" and "diagnosis," conjuring up a name that would in time become infamous: Theranos.

She quickly became acquainted with the venture capitalists keen to invest in the next hottest property in medical devices. By year's end Theranos had secured over $6 million in funding, seducing investors to part with $92 million by 2010, despite the company's operating in stealth, devoid of even a website. This was by design. Holmes cultivated the image of a technology visionary, idolizing the style of Steve Jobs, even down to the turtleneck garb. Like Jobs, she insisted on the highest level of secrecy, largely forbidding employees from discussing what they were working on, even with one another. Every decision, no matter how trivial, crossed her desk. Despite this secrecy, the money flowed in from investors drawn to the whispered potency of an alluring idea: a simple test that could diagnose a range of conditions from a meager few drops of blood, promising to confine the bane of needles to the past.

Holmes assured investors that the tests were rapid and highly accurate—from just a single, tiny drop of blood, Theranos promised diagnostics on dozens of conditions. Luminary figures flocked to the board of directors, including political titans Henry Kissinger and William Perry. As the profile of the nascent company grew, the much-lauded Holmes metamorphosized into the central deity of a personality cult. Money and prestige continued to roll in. As her personal star grew, the popular press babbled with an abundance of hagiographic depictions. Holmes's confident demeanor and promises to fundamentally alter the diagnostic industry enthralled the press, and cover stories followed in *Forbes, Fortune,* the *Wall Street Journal,* and *Inc.*, which referred to her as "the next Steve Jobs."

By 2014, Theranos was valued at $9 billion. As Holmes held a 50 percent stake, *Forbes* pinned her net worth at a cool $4.5 billion, declaring her the world's youngest self-made billionaire. And not only was Holmes promising to disrupt the diagnostic field, but she did so under a mantle of consumer choice, partnering with drug store chain Walgreens to offer their blood tests on site. Petty legal impediments to such expansion were quickly demolished—a 2015 Arizona bill coauthored by Holmes decreed that patients could now order blood tests without a doctor's input. Holmes enthused that the new law was "just about having the access to your own health." The more astute observers, however, noticed that the law meant significant financial gain on the horizon for Theranos, which would analyze these tests with the wondrous device at the heart of their ambitions: the Edison machine.

Despite all the effusive praise and phenomenal money invested in this technology, the scientific community made no secret of their substantial doubts. Theranos refused to release any details of their ostensibly revolutionary tests, claiming that to do so would undermine its business. But to scientists, this excuse rang hollow. A 2015 *Journal of the American Medical Association* editorial by John Ioannidis criticized the nebulous nature of "stealth testing," expressing concern about the rationale behind promoting widescale diagnostic testing. In his opinion, "The main motive appears to be to develop products and services, rather than report new discoveries as research scholarship." Nor was Ioannidis alone in his skepticism, with numerous other scientists also expressing reservations. This was reinforced by a bombshell revelation that results from Edison were so unreliable that Theranos was using competitors' machines. As if to mark the extent of Holmes's trouble, a damning investigative report by Pulitzer Prize–winning author John Carreyrou appeared in the *Wall Street Journal* just months after the same publication had lauded her.

Theranos took a combative stance in response, admonishing the article as terrible journalism, fueled by disgruntled employees. But such bluster fell far short of a convincing refutation and, over the course of a few weeks, the trickle of woes became a torrent. In January 2016 the Centers for Medicare and Medicaid Services (CMS) sent Theranos the results of their investigation into the company's laboratory facilities, uncovering alarming inaccuracies in their tests likely to cause "immediate jeopardy to patient health and safety." Sanctions were imposed later that year, which forbade Holmes from owning or operating a laboratory for at least two years. Other investigations cast further doubt on the results from the Edison machine, forcing Theranos to void a large number of tests. Former partners Walgreen quickly abandoned the company and proceeded to sue for $140 million in damages arising from breach of contract.

Criminal investigations began too, based on considerable evidence that Theranos had misled government regulators and investors over the accuracy of their devices. In a distinct reversal of its fortunes, the company that had successfully changed Arizona law found itself the subject of a lawsuit filed by the Arizona attorney general over their "long-running scheme of deceptive acts and misrepresentations" related to the company's blood-testing equipment. Layoffs ensued, and Theranos labs closed rapidly, failing every ensuing inspection. By June 2016, *Forbes* had drastically reevaluated the net worth of Theranos, and with it Holmes, to a value they felt was more reflective of reality: $0, absolutely nothing.

The Edison was nothing more than an elaborate Mechanical Turk,* feigning the appearance of rigor with an elaborate parlor trick. Much has been written of the saga, with all its dishonesty, ineptitude, and hubris,† but there is something important underpinning this story that should resonate with us. Theranos's stratospheric rise was due in large part to the incredible sums of money it was able to raise, but alarm bells should have sounded much earlier. The most obvious warning sign was the promise of accurate tests on microscopic quantities of blood; there are well-established chemical and physical reasons why tiny drops of blood are difficult to work with, and indeed much of the skepticism focuses on this aspect. Still, this wasn't completely insurmountable—it was possible the Edison technology had somehow made a huge leap in microfluidics and that investors had been assured of this.

But there was a subtler reason why the claims of Theranos should have rung alarm bells, and one far more fatal. For all the supposed savvy of high-technology investors, a three-minute conversation with a statistician should have been enough to convince them to steer clear. While Theranos tried to paint its shotgun approach to

---

* The Mechanical Turk was a machine built in 1770 that ostensibly played chess. For nearly 84 years, it beat figures from Napoleon to Benjamin Franklin. But it was really an elaborate hoax, concealing a hidden player. At various times, these hidden operators included some of the world's finest chess masters.

† Movie rights to the Theranos saga have already been sold.

diagnostics as a virtue, this idea is inherently doomed to failure. Why is this? Well, first we need to acknowledge that medical tests alone rarely produce a smoking gun but are ordered when other signs appear that might indicate a condition. The allure of screening for disease before symptoms appear is undoubtedly appealing, but often it might be medically useless, and in the absence of symptoms such tests can be positively misleading at best and actively harmful at worst.

To understand why this is, it is useful to introduce a pair of important concepts. The first of these is *sensitivity*, which is a measurement of how many positive results are correctly identified as such. For this reason it is sometimes called the "true positive rate"; if 100 people had a certain disease, and a test correctly identified 90 of them, then the test has a sensitivity of 90 percent. The converse of this is *specificity*, which is the proportion of negatives that are true negatives—also known as the "true negative rate." In a perfect world, a test would be both 100 percent sensitive (only catching true positive results) and 100 percent specific (completely avoiding false negatives). Alas, this is not the world we inhabit—even high-quality tests fall short of this mark; in practice, tests with above 90 percent sensitivity and specificity are considered good tests. Crucially, they do not yield certainty in isolation, rendering their unqualified use suspect in the extreme.

We've encountered an example of this before when we looked at HIV testing, which has virtually perfect specificity, rendering it unlikely to give a false negative. But even with an extraordinarily high sensitivity of around 99.99 percent, we've seen already how 50 percent of positives in the low-risk cohort are false positives. The diagnostic power of any test is intrinsically related to both its sensitivity and specificity, and this must be carefully interpreted. To complicate matters further, while these parameters are independent of disease prevalence, the positive or negative prediction value of a test does depend on how common the illness is, and to compute it requires careful application of Bayes's theorem. A scattershot approach to diagnostics without taking stock of other factors is intrinsically flawed. In a scathing editorial, Eleftherios P. Diamandis, a Canadian clinical biochemist, laid the problem bare:

A lay person whose PSA is 20 µg/L will assume, based on statistics, that he would have a more than 50 percent chance of harboring prostate cancer, and ask for a biopsy. However, if his PSA a few days earlier was 1 µg/L, his chances of having cancer are virtually zero, the likely cause of his PSA increase being acute prostatitis, a benign and treatable condition. A male with a positive "pregnancy test" will likely be totally confused but a trained physician would look for testicular cancer.

Holmes made "democratizing health care" a central plank of her advocacy, inviting patients to self-test—but, in doing so, she ignored the solid rationale behind physicians ordering specific tests, and for limiting the population to be screened. The wider the diagnostic net is cast, the greater the rate of false positives. The much-vaunted promises by Theranos that they would test the same drop of blood for up to 30 conditions only made things worse—when multiple independent tests are conducted, the testing flaws are exacerbated. To illustrate this, if every test had a 90 percent specificity, then the odds of getting at least one false positive after 30 tests would stand at an alarming 95 percent. And even if we were able to make each of these tests 99 percent specific, at least one false positive would occur more than 25 percent of the time. This is an inherent limitation of multiple independent tests, with each extra test decreasing the net prediction accuracy into nothing more than noise.

Even if the wonder machine had actually performed its function, its promises were utterly undermined by statistical reality. The idea that one can simply throw a barrage of tests at a wide cohort of patients, with no a priori information, and use these to divine their health status is completely irrational. The Edison machine, far from liberating its users from the need to visit doctors, would have made them slaves to needless fear. The finger of blame for misleading patients, lawmakers, and investors has been firmly thrust in the direction of Elizabeth Holmes, and there is no doubt many of her claims bordered on the fraudulent. Nor is there any doubt that her attempts to counter criticism devolved into outright obfuscation. Blaming Holmes entirely for the Theranos debacle, however, would be misguided. Had the investors done their due diligence to ask basic questions about her claims, it is doubtful they would have become quite so enamored. There is an adage that is especially pertinent here: Fools and their money are soon parted.

14

SIZE MATTERS

*How to Properly Assess the Strength of Evidence in Health Claims,
from Absolute Risk to Statistical Significance*

In October 2015, carnivores around the world were greeted by an unwelcome finding: Processed meat was carcinogenic. The *Daily Express* screamed BACON AND HOT DOGS CAUSE CANCER–AND ARE ALMOST AS BAD AS SMOKING. Not to be outdone, *The Guardian* proclaimed that PROCESSED MEATS POSE SAME CANCER RISK AS SMOKING AND ASBESTOS. These headlines stemmed from an arresting press release from the International Agency for Research on Cancer (IARC), an arm of the WHO tasked with researching the causes of cancer. They had announced that processed meat increased the risk of bowel cancer by almost 18 percent, classifying it as a Group 1 carcinogen alongside smoking and radiation. The same communiqué classified red meat as Group 2A, defined as "probably carcinogenic" to humans. The idea that meat was as dangerous as smoking caused widespread consternation.

These dire headlines, however, were abject nonsense. IARC's arcane classification system is based not on *degree* of risk, but on the *strength of evidence* for that risk. This means that something resulting in a tenfold increase in cancers would receive the same classification as something that only increased risk a negligible amount. The classifications do not convey how dangerous something might be—only our certainty that it might be dangerous. Group 1 agents are those for which there is strong evidence of risk, and includes smoking, sunlight, and alcohol. Group 2A and 2B respectively are "probably" and "possibly" cancer-causing. This in practice translates as limited or ambiguous evidence of risk. Given the philosophical difficulties of proving a negative, Group 2 is something of an epidemiological dumping ground. As of 2018, the only recognized Group 4 agent (probably not carcinogenic to humans) is caprolactam, used in the manufacture of yoga pants, among other synthetic products.

If all this sounds unbelievably obtuse and counterintuitive, it's because it is. As a scientist working in cancer research, I understand the rationale behind stratifying risks. As someone who communicates science to the public, I curse the lack of foresight behind a classification system rife with potential for confusion. When a layperson is told everything from shift work to coffee is "possibly carcinogenic," it's understandable they don't interpret this as "evidence of risk is weak and unclear." As science writer Ed Yong noted, the IARC "is notable for two things. First, they're meant to carefully assess whether things cause cancer, from pesticides to sunlight, and to provide the definitive word on those possible risks. Second, they are terrible at communicating their findings." Criticism of clarity aside, just how dangerous is processed meat?

To answer that question, first we need to look at the underlying data. In the UK, 61 people per 1,000 develop bowel cancer during their lifetime. For those who ate the least amount of processed meat, the rate was 56 per 1,000, whereas for the heaviest consumers it was 66 per 1,000. Among the most passionate carnivores, there were 10 more bowel cancers per 1,000 people than the group that abstained. *Relative risk* increase is defined as the increase in risk of the exposed group, relative to the unexposed group. Here, that's (66–56)/56, or 10/56—roughly 18 percent, the figure IARC quoted in their press release. Another way of looking at this is in terms of *absolute risk*. The difference between the lifetime risk of bowel cancer for processed meat eaters and non–meat eaters is 10/1,000, which is exactly 1 percent. In effect, the risk of getting bowel cancer through one's lifetime is 1 percent higher for heavy consumers of processed meat than it is for those who eat absolutely none. This latter figure is undeniably a lot less frightening.

The way in which probabilistic data is reported has a huge effect on how we understand it, and on our emotional processing of that information. This is especially true of information pertaining to our health and mortality; astute observers of the media may have noticed an ongoing crusade by many tabloids (and several respectable newspapers who should know better) to reduce the entirety of creation into a neat cures/causes cancer dichotomy. Relative statistics always sound more stark than absolute numbers, despite reporting the same information. As the more sensational-sounding figure will be the relative risk, it is far more likely to be co-opted by media outlets. It is, however, likely to mislead, and there is good evidence that absolute risks are better understood by the public.

Media outlets and WHO bodies aren't the only ones guilty of overreliance on relative risk. This class of statistical overcompensation is glaringly obvious in the pharmaceutical sector, where drug companies are prone to quote the effectiveness of their drugs in relative terms to instill the impression of a more effective product.* For example, imagine a trial with 2,000 patients with heart conditions, 1,000 of whom are given a placebo and the other 1,000 of whom are given a new drug. If there were five heart attacks in one year in the placebo group, and only four in the drug group, then absolute risk reduction is only 1/1,000, or 0.1 percent. This isn't especially impressive; assuming the difference between the groups isn't just a happy fluke, doctors would have to give the drug to 1,000 people to prevent a single heart attack. Given the sheer cost of bringing new drugs to market, the better-looking 20 percent relative-risk figure will most likely be embraced.

A variant of this is often seen in economics and politics, where statistics are deployed for the purposes of fallacious comparisons. If a house valued at $200,000 falls in value by 50 percent in one year and rises by 50 percent the next, it may be reported that the house has recovered its former market value. Yet this is patently false—at the end of the first year, the house is worth only $100,000. Increasing by 50 percent the next year, it rises to $150,000, a figure only 75 percent of its initial value. This occurs because the 50 percent statistic was

---

* Ben Goldacre's book *Bad Pharma* explores the problems of the pharmaceutical companies' conduct and trial reporting.

relative to two different baselines. The first was the initial value, and the second the depreciated value. The vital point is that percentages can't simply be added and subtracted without cognizance of the problem because they often are relative to different figures.

There is one avenue of equivocation I have deliberately avoided until now: the thorny question of *statistical significance*. We frequently encounter headlines alerting us that something once thought benign has a statistically significant link with cancer, or claims that a certain diet can reduce one's risk of dementia by a statistically significant degree. But what precisely does this mean? Significance is perhaps one of the most misunderstood words in all of science, occasionally even by scientists themselves. Imagine that we've created a new wonder drug we believe will benefit migraine sufferers. Our hypothesis is that this drug, agent X, reduces migraine frequency. In contrast, we have a null hypothesis, a default position that asserts there is no relationship between agent X and migraine frequency. We undertake an experiment, dividing our subjects into two groups. One of these groups is given the agent, designated the experimental arm. The other group—the control arm—is given a placebo. When the experiment concludes, the real question we want to answer is this: Did agent X really have an effect, and can we reject the null hypothesis?

Answering this in practice requires statistical methods. People are incredibly diverse—both arms will have a distribution of patients with differing responses. In a perfect world, our sample group would be perfectly representative of reality, but because we only have a finite number of subjects, this isn't possible. Outliers in either or both groups might skew the average, potentially misleading investigators. Both arms are likely to differ to some extent by chance anyway, so to determine whether true differences exist, we employ statistical tools. When applied correctly to well-conducted experiments, they are invaluable at cutting through the noise and determining whether there are real differences between the arms. A result is deemed statistically significant when it is considered unlikely to have arisen by chance, implying that the result is a real one. Importantly, statistical significance merely implies the drug has some impact; it doesn't necessarily mean that the impact is particularly substantial, which the commonplace understanding of significant often implies.

But if these procedures are followed, why do so many purported links turn out to be equivocal or wrong? Often the fault lies with scientists and physicians, who are not immune to the errors we've previously encountered. While reputable scientific papers go through meticulous peer review, statistically dubious assertions can and do slip through the cracks. Naturopathy, a branch of alternative medicine, is a prime example of this. Naturopathic disciplines—which encompass reflexology, homeopathy, and cranial therapy—are based on vitalism, the notion that some ethereal life force is responsible for sickness and health. This idea has long been refuted by empirical findings, and there is no reliable evidence that any of these treatments have any medical benefit whatsoever. Yet, even in this scientific era, they remain curiously popular. Part of the reason for this is undoubtedly due to an appeal to nature and the mistaken perception such interventions have no side effects.* By offering a simple formula for health with easy answers, naturopathy downplays the complexity of both medicine and our very bodies.

But curiously, naturopaths insist that there is scientific evidence showing their therapies to have a statistically significant effect on patient outcome. However, if these treatments have neither plausible mechanism nor clinical effect, how exactly can these mutually exclusive statements coexist? The answer lies in the subtle nature of statistical significance. Statistical approaches are enlightening if the data is of sufficiently good quality and the analysis appropriate to that situation. But employed haphazardly, results become meaningless. The handful of positive studies naturopaths so dearly cling to are invariably of low quality, performed on small sample groups.

* On homeopathy, comedian and former physicist Dara Ó Briain has an amusing take: "The great thing about homeopathy is that you can't overdose on it. Well, you could fucking drown."

This is important because in small groups a single outlier skews the entire analysis, and the smaller a sample group is, the less robust the conclusions drawn. Tellingly, apparent benefits disappear when larger groups are analyzed and trial quality increases, as one would expect. The much-touted significance of the intervention effect is entirely illusory.

Perceived benefits reported are placebo effects, perhaps more accurately considered the consequence of *regression toward the mean.** This is the observation that when a measurement of a variable is extreme in the first instance, the next measurement tends to be closer to average. For example, people usually seek help when their symptoms are at their zenith. This is an extreme state, and over the passage of time recedes to a more normal baseline. But many still attribute their recoveries to long-debunked folk medicine rather than consider the phenomenal talents of their own immune system. Nobel laureate Peter Medawar observed that: "If a person is (a) poorly, (b) receives treatment intended to make him better, and (c) gets better, then no power of reasoning known to medical science can convince him that it may not have been the treatment that restored his health."

This illustrates an underappreciated aspect of scientific research: not all studies are created equal. When statistical significance is found this does not always mean an effect is present. Sadly, meaningless significance studies blight many avenues of research where statistical analysis is to the fore, including medicine and genetics. In 2005, John Ioannidis published the provocatively titled paper "Why Most Published Research Findings Are False," which drew some arresting conclusions. In medical fields, many significant results are simply artifacts of poor trial design, underpowered studies, or groups with too few participants to draw meaningful conclusions. In his work, Ioannidis outlined six indicators that should be remembered when evaluating the veracity of any claim:

---

* There is solid evidence that the placebo effect is relatively small, and that regression toward the mean accounts for much of the reported benefits of sham interventions.

1. *The smaller the studies conducted in a scientific field, the less likely the research findings are to be true.* If the sample is small, the chances of the group being representative is lower and the rate of false positives increases. This is precisely the kind of studies that advocates of naturopathy cling to, with small sample groups and poor-quality construction.

2. *The smaller the effect sizes in a scientific field, the less likely the research findings are to be true.* Correlation itself is important, but effect size matters too. Effect size is a measure of how strong the phenomenon is, useful for determining whether the observed relationship is mere chance or something more substantial. If the effect size is tiny, effects may be nothing more than chance.

3. *The greater the number and the fewer the selection of tested relationships in a scientific field, the less likely the research findings are to be true.* Simply put, if an experiment generates lots of possible relationships, then by chance alone some of these might be false positives. With lots of possible correlations to examine, it is too easy to cherry-pick those which might by chance alone show a possible statistical connection.

4. *The greater the flexibility in designs, definitions, outcomes, and analytical modes in a scientific field, the less likely the research findings are to be true.* If one allows more leeway in definitions, bias can creep in and a "negative" result can deftly be manipulated into a false positive one.

5. *The greater the financial and other interests and prejudices in a scientific field, the less likely the research findings are to be true.* In the biomedical field especially, conflicts of interest often arise between funders and results, inviting bias. As Ioannidis makes clear, the conflict of interest does not have to be financial; scientists are not immune to ideological devotion to certain ideas, and this can alter results.

6. *The hotter a scientific field (with more scientific teams involved), the less likely the research findings are to be true.* This is a counterintuitive but important observation. While more investigation of a certain area should, in principle, increase the quality of the findings, the opposite occurs when groups compete aggressively. In such cases, time becomes of the essence, and research teams might be inclined to publish prematurely, leading to an excess of false positive results. Ioannidis and colleagues termed this phase of research the "Proteus phenomenon," capturing the rapid alternation between extreme research claims and equally extreme refutations.

These worrying and meticulously researched observations raise an urgent question: If most published research findings are wrong, then what use is scientific inquiry? How can research have any meaning? The first thing to note is that the kind of research Ioannidis refers to is not "all" research, but rather those studies pivoting on the "ill-founded strategy of claiming conclusive research findings solely on the basis of a single study assessed by formal statistical significance, typically for a p-value less than 0.05." This is undoubtedly a problem in fields that rely heavily on statistical correlation alone, a scattershot approach to scientific endeavor. But it is markedly less of a problem when experiments are well planned, based on known principles. Events recorded by the Large Hadron Collider, for example, undergo stringent statistical analysis to determine whether new fundamental particles have been detected. The gold-standard threshold for statistical significance in particle physics is so extraordinarily high that false positives are vanishingly unlikely.

In parts of medicine and biomedical science, however, the problems Ioannidis describes are ubiquitous. In these fields, complex interactions are difficult to escape so researchers resort to "discovery-orientated" exploratory research, rather than starting with a well-formed hypothesis. This lends itself to false discoveries, with chance results gaining precedence they simply don't deserve. Part of the problem lies in the arbitrary nature of where one sets the "cutoff" value for significance, often called the p-value. A p-value of less than 0.05 is often taken to mean a result is significant, and many researchers are slavishly devoted to this number. The problem is that it was never supposed to be a true measure of quality or even an ideal. Biologist Ronald Fisher pioneered it in the 1920s as a statistical rule of thumb, an informal test to determine whether a result was worthy of a second look.*

At that time, a movement for mathematical rigor in statistics was gaining momentum, led by Fisher's archrivals, Polish mathematician Jerzy Neyman and English statistician Egon Pearson. Neyman and Pearson formalized concepts such as statistical power but held Fisher's innovations in contempt. Neyman dismissed some of Fisher's innovations as "mathematically worse than useless," while Fisher scorned Neyman's approaches as "horrifying for intellectual freedom." Other statisticians grew tired of the feuding between these pioneers and simply melded their frameworks together. Fisher's rule of thumb was forced into Neyman and Pearson's mathematical framework, elevated to something it was never supposed to be.

This in turn has let to abuses and misunderstandings; some researchers effectively data-mine, haphazardly seeking statistically significant relationships without due consideration of whether this is truly meaningful or merely a product of chance. David Colquhoun, a fellow of the Royal Society, has long castigated those who engage in such practices, memorably stating that "the function of significance tests is to prevent you from making a fool of yourself, and not to make unpublishable results publishable." Because of the fitting acronym it yields, the term "Statistical Hypothesis Inference

---

* As this section is already quite technical, I've played fast and loose with the concept of a p-value, defined here as a general test to ascertain whether a result might have some property that makes it worthwhile to investigate further.

Testing" has been suggested for this type of data-mining. Without a mechanism of action or solid underlying principle, correlations should be treated with caution. Simply throwing lots of post hoc tests at the data in the hope of finding significance will usually yield a result, though it will usually be meaningless rather than enlightening. As economist Ronald Coase once observed: "If you torture the data long enough, it will confess." Such confessions, of course, are not likely to be reliable.

So why then do some scientists publish results that are underpowered and questionable? Partly because statistical ineptitude is not just a condition that affects nonscientists. But another factor stems from more depressing motivators: publication bias and pressures on scientists. Scientific journals are much less likely to deem negative results worth publishing, which places researchers under immense pressure to find links between phenomena at the risk of these links being spurious. This is profoundly short-sighted. Null results are every bit as valuable to our understanding as significant findings. It is far more useful to know that a drug doesn't work, for example, than to be presented with incorrect assertions that it does.

Compounding this, scientific enterprise has of late been infected by a destructive "publish or perish" dictum, where funding isn't forthcoming if scientists are not deemed to be producing enough positive results—a rewarding of quantity over quality that imperils all of us. For all these reasons, one must be wary of single studies, especially in the avenues of medicine and other fields where correlation rather than mechanism is observed. A statistically significant result does not by itself mean the result is "real," a vital caveat we mustn't forget.

Incidentally, John Ioannidis and I have worked together on this problem before, modeling the impact of "publish or perish" pressure on the trustworthiness of published science. Predictably perhaps, our results suggested that the current paradigm tends to reward dubious results over rigorous inquiry, thereby perpetuating the problem. Science thrives on reproducibility, and without it a result simply doesn't stand. Accordingly, this issue has become more keenly discussed in recent years. It has motivated the Open Access and Open Data movements, where scientists are encouraged to submit all results—positive or negative—as well as the data used to support conclusions so other researchers may use it.

There are also powerful tools for comparing across several studies, especially where results conflict or the quality and power of those studies vary. One such method is *meta-analysis*, which can be thought of as a study of all studies, gauging them for quality and facilitating a clearer picture of all the available data. This crucially requires an abundance of studies to weigh in terms of quality and scope, and this is precisely why the results of any single study should always be taken as preliminary and subject to change. The findings of science are always provisional and ever in flux. This is not a weakness but the vital core of self-correction upon which science relies.

In these chapters, we've encountered some examples of how statistics and numbers can confound us and how bumps in logic can render them completely opaque and potentially misleading. Naked numbers devoid of context can convey misleading impressions, even if accurately reported, and it can take some finesse and clever questions to see the true message they embody. Statistics are a powerful tool, but our collective interpretation of them often leaves much to be desired. If we are to truly benefit from statistical analysis, we owe it to ourselves to improve our understanding lest we fall victim to misconception.

The abuse of statistics can also be the best friend of a demagogue, allowing perverse misrepresentation to slide into argument unimpeded, cloaked in the mystifying armor of numeric confidence. One need only look at political discourse where baying politicians lob context-free numbers at one another like argumentative grenades to score points, oblivious of and ultimately unconcerned with the interpretation or veracity of the figures. Such spectacle is depressing, but we might reasonably ask: How can this be avoided?

On an individual level, there is no substitute for becoming aware of the uses and abuses of statistics. As a society, our collective fear of numbers should make us wary of statistics wielded as absolutes. We lack confidence in our numerical ability, and so abuses perpetuate unchallenged. Yet the basics, such as those covered in this chapter, are readily graspable. One does not need to be an expert to spot the more alarming flaws prevalent in common usage.

There is also compelling evidence that statistics are better understood when presented in real numbers—a technique known as natural frequency reporting. For example, if patients are told that a medication has a 10 percent chance of a certain side effect, they are more likely to have a context for this figure if told that "in a group of 100 patients on this medication, we'd expect 10 to have this side effect during treatment." Even professionals benefit from natural frequency reporting; while a shocking number of medical professionals incorrectly calculated the odds of a patient having HIV, as in our earlier Bayes's theorem example, this number dropped dramatically when the situation was reported in natural frequency, as laid out in the branching tree examples in chapter 12. When presented this way, doctors surveyed almost unanimously got the correct result—a complete inversion of the case where the numbers had been reported as statistics.

The vital point to garner from all this is the reality that, while statistics have a certain intuitive appeal, they mask reams of subtlety and complexity that can thoroughly bamboozle, sending us scurrying to falsehoods. Reinterpreting these numbers in the context of what they actually tell us is a step we too often miss. When in doubt, it pays to ask the somewhat deeper question of what a sound-bite statistic actually means, and what we can infer from it. Without this acid test, we risk being led astray by the numerically confused or the ideologically perverted.

Bald numbers devoid of all caveats may tell us less than nothing, but too often they are fuel to the fire of sensationalism. But it's not solely the statistics themselves that mislead us; it can be the trust we place in the sources from whence they arise. The narrative in which they're delivered often shapes our perception. And as the media we consume is our primary source for the myriad numbers we're confronted with every day, this influence cannot be denied. The role of both traditional and emerging media on our understanding of the world around us is something we need to grasp if we're to appreciate how easily we can be misled—and how to circumvent this.

PART 5

News of the World
How Media Indulges Bad Thinking

"Newspapers are unable, seemingly, to
discriminate between a bicycle accident and
the collapse of civilization."

—GEORGE BERNARD SHAW

15

SKEWING THE BALANCE

How Donald Trump and Creationists Get Prime Time:
False Balance and Manufactured Controversies

When the dust finally settles and future historians look back to the early twenty-first century, the bizarre events of the 2016 US presidential election will still hold grotesque fascination. The contest between Democratic candidate Hillary Clinton and the Republican nominee Donald Trump was anything but typical. I was in Florida on November 8, 2016, watching the count in a bar in Ybor City with several fellow scientists. Like most people worldwide, we expected to see America electing its first-ever female president. After all, she was the undoubted favorite, cutting a much less controversial figure than her opponent. But as Florida was called for Trump, the uneasy sensation that we were witnessing an unprecedented electoral upset grew stronger. By early morning, the result was all but confirmed. Contrary to predictions, Trump had won the presidency.

The shock waves this sent across the world will no doubt captivate political scholars for decades. While it is still too early to know the long-term consequences of this event, a cautionary tale is already evident. Clinton was a relatively conventional selection, with ample state experience. In the words of outgoing President Barack Obama, there had "never been a man or a woman more qualified than Hillary Clinton to serve as president of the United States of America." Despite that accolade, her campaign was not devoid of flaws and she was dogged by controversy over her use of a private email server during her time as secretary of state. Yet there was no doubt Clinton would abide by norms both constitutional and political, as had virtually all major presidential candidates before her.

Trump, however, was no ordinary candidate, displaying obvious contempt for such norms. A reality TV star and businessman of dubious skill, his intention to run for president initially drew bemused reactions across the political spectrum. The expectation was quite simply that Trump was an amusing sideshow, a man given to narcissistic braggadocio, devoid of political insight or substance. His attacks on political opponents lacked even a modicum of decorum, with insults and smears employed against foes on both sides of the aisle.* Yet to the surprise of almost everyone, and to the chagrin of many, Trump managed to seize the Republican nomination.

From the outset, the campaign was a huge departure from American political norms. Running on an explicitly racist platform, he directed ire at Muslims, Hispanics, and people of color. He had no qualms about deploying misogynistic language, slamming women who irked him as "bimbos" and "fat pigs," and he remained unconcerned with the growing number of accusations of sexual assault against him. Evidence of duplicitous business practices did not sink him, nor did the bizarre spectacle of his being lauded by openly racist organizations such as the KKK, the American Nazi Party, and the nascent "alt-right" movement. Even allegations of his possibly treasonous associations with Russian agencies bent on interfering in the election cycle could not dent him. To a traditional candidate, any of these transgressions would have dealt a fatal blow—but instead, Trump emerged unscathed from an ever-increasing series of scandals.

With Trump's unexpected ascent, news organizations scrambled to cover the election in as even-handed a manner as possible. Under normal circumstances, media outlets tend to treat candidates as approximately similar, gauging their strengths, comparing their flaws, and maintaining impartiality, thus framing the election as a choice between two roughly comparable options, bound by the same rules of engagement and accountable to the same standards. But Trump refused to be bound by such convention—his attacks grew ever more personal and his lies ever more outrageous. The Pulitzer Prize–winning fact-checking PolitiFact awarded him the 2015 "Lie of the Year" award, but this proved no deterrent. Trump

---

* Trump was also the most vocal of all the Obama birthers whom we encountered in an earlier chapter.

simply lied at an even greater rate, sending media outlets scrambling to cover his latest utterances.

These utterances came thick and fast, prompting *The Guardian*'s US correspondent Alan Yuhas to note that "Trump lies like he tweets: erratically, at all hours, sometimes in malice and sometimes in self-contradiction, and sometimes without any apparent purpose at all." Even when these falsehoods were transparent, media organizations felt obliged to report his accusations, bereft of evidence as they were. But this proved an insidious strategy, as it amplified his accusation-filled ramblings, propagating toxic ideas to a receptive audience. Attempts to debunk these claims seemingly fell on deaf ears, and media commentators belatedly began to despair of the rise of "post-truth" politics, an emotion-driven culture where factual rebuttals of incendiary claims can simply be ignored.

In their desperate attempt to impose normality on a situation that was anything but typical, the media gave copious airtime to baseless accusations. Worse again, attempts at even-handedness pivoted on the delusion that Clinton and Trump were equivalent candidates with comparable flaws. With a misguided understanding of impartiality, outlets presented Hillary's relatively minor scandals as being on an equal scale to Trump's astounding transgressions. This played to his advantage, and he pushed the illusion of "Crooked Hillary." Too late into the election cycle, media outlets began to realize that they had effectively normalized something bizarre. In trying to treat the grotesque spectacle of Trump's ravings as a typical political campaign and a mirror of Clinton's, they had inadvertently given his awful pronouncements a veneer of legitimacy.

It was a fool's errand to treat Clinton and Trump as symmetrical candidates when they were not remotely comparable. Yet the damage was done—Trump had exploited the media's willingness to present him as a normal candidate and, as much as this benefited him, it hurt Clinton. By September 2016, Paul Krugman, an economist and a *New York Times* opinion columnist, took his colleagues to task for their insipid performance:

If Donald Trump becomes president, the news media will bear a
large share of the blame. I know some (many) journalists are busy
denying responsibility, but this is absurd, and I think they know
it. As Nick Kristof says, polls showing that the public considers
Hillary Clinton, a minor fibber at most, less trustworthy than a
pathological liar is prima facie evidence of massive media failure.

The error media outlets made is a classic one known as *false balance*. This occurs when one tries to treat two opposing positions
as equally supported by evidence when they are not. It frequently
transpires when a shared trait between two subjects is wrongly
taken to imply equivalence. This is like arguing that there is no
difference in keeping a house cat or a tiger as a pet as they're both
just felines. If one position or claim is supported by an abundance of
evidence while another is bereft of supporting data, it is fundamentally flawed to equate them solely on the basis that they are mutually
opposed positions. This opposition alone does not inherently make
them equally worthy of consideration—but too frequently this
subtlety is missed, presenting a vacuum of logic that can be readily
exploited by the foolish or nefarious.

This manifests most obviously in coverage of contentious
issues and debates. Respectable media outlets pride themselves
on avoiding bias or partisanship. This is a laudable position to
take, as robust debate in a healthy society is vital. We all have a
predisposition toward partisan sources, and informed discussions
can steer us out of damaging echo chambers. For conscientious
editors, broadcasters, and writers, objectivity is an ideal to strive
for. But impartiality should never mean false equivalence. When
the weight of evidence points incontrovertibly in one direction,
doggedly reporting both "sides" as equally valid lends an air of
respectability to terrible ideas and nonsense claims. False balance
arises when one attempts to present opposing views as being more
equal than the evidence allows. However, if the evidence for a
position is virtually incontrovertible, it is profoundly mistaken

to treat a conflicting view as equally legitimate and worthy of consideration.

False balance thrives on the illusion of equality between conflicting claims, with no true regard to the evidence behind them. It is easily manipulated, even on objective scientific topics. Take, for example, a subject like alternative medicine: While there is an utter paucity of evidence of efficacy, the fact that some patients report alternative medicine as being beneficial is taken as valid evidence equal to the scientific trials and studies that show no effect beyond placebo. This is ludicrous, as the a priori assumption that good journalism requires mutually opposed views to be treated as equally valid simply doesn't hold when the overwhelming weight of evidence points resolutely in one direction. It does, however, require a certain amount of expertise to gauge this, and for media outlets, discerning between valid science and pseudoscience can be a difficult task.

Although no malice or bias is intended, the net result of such ineptitude is encountered all too frequently. As we saw, Andrew Wakefield's falsehoods over the MMR vaccine cost innocent lives, but this would have been impossible without the inadvertent cooperation of media outlets who posed his fearmongering as equivalent to the overwhelming body of scientific evidence that exposed it as nonsense. None of this is intended to lay the blame for these deleterious and even fatal consequences at the feet of media outlets alone, but the very fact that anti-vaccine activists could exploit the ideal of impartiality to circulate baseless scare stories unimpeded is a worrying one.

We have learned precious little from this debacle; false balance still crops up on scientific issues with depressing regularity, even on the subject of vaccination, which is still too often treated as a contentious topic. Even Wakefield, architect of the MMR panic, has not been completely expunged from the pulpit. In 2016, he was at the center of controversy again, promoting a documentary that claimed the Centers for Disease Control and Prevention was covering up vaccine damage. It was even included in the Tribeca Film Festival at the behest of Robert De Niro, drawing intense criticism.

Alas, criticism is irrelevant because, with false balance, no publicity is bad publicity; it is merely a platform to push a narrative.

One regional Irish radio station asked me to debate with Wakefield himself. I urged them not to give him a platform at all, outlining why this would be misguided. The producer cited strong local interest, claiming a rival broadcaster had offered him a slot with a sympathetic host and no counterpoint. The ultimatum was simple: He was getting a slot either way; the only question was whether he would encounter opposition or not. I grudgingly agreed to appear, with the caveat that I would articulate why giving him a platform to air discredited views was itself misguided. The experience was immensely frustrating. I outlined why Wakefield's claims were meritless and, predictably enough, he rattled off claims of a monstrous conspiracy, accusing me of being part of it. It culminated in a string of ridiculous assertions from an irate Wakefield, before I terminated the interview. The segment that aired was whittled down substantially, resulting in a disconnected mess, and my caveat on false balance was absent completely.

Disheartening as the experience was, it was also a valuable lesson: No matter how noble the intentions, presenting science and pseudoscience equally in an adversarial format gives the false impression that an issue is scientifically contentious. It allows empty claims to leech vampirically off the legitimacy of well-established theories, and provides free rein for dubious motivations to masquerade as scientific opinion, enabling the cynical to manipulate the naive.*

---

* In 2017, Wakefield was due to receive an award from a homeopathic society in Regent's College London and screen his movie, which we protested against. Approached for comment from *The Telegraph*, I didn't hold back on the issue of false balance, stating: "Wakefield is a long-debunked fear merchant whose attempt to paint himself as a Galileo-like figure is at once completely narcissistic and utterly dishonest. Whether by oversight or intention, giving Mr Wakefield a platform on vaccines is a grievous mistake, given that we're still reeling from the damage his falsehoods inflicted on public health. Not only are his claims devoid of evidence, they are vividly disproven by the overwhelming scientific data to date. When the evidence points in only one direction there is no debate, yet by hosting someone so notorious [the university] gives the perception his assertions might have merit. They do not." The screening was ultimately canceled.

False balance isn't solely confined to media issues, however. In comparatively recent history, lung cancers were a rarity; in 1878, they accounted for less than 1 percent of cancer incidences. So rare were malignant growths in the lungs that surgeons took special notice when they occurred, viewing them as once-in-a-career oddities. Yet, by the early part of the twentieth century, lung cancer rates had risen dramatically, hitting 10 percent of cancers by 1918 and 14 percent by 1927. Various explanations were posited, ranging from increased air pollution to the environmental fallout of the First World War, but none of these proposed explanations could truly account for the data, which showed a marked increase in lung cancer incidence, even in countries unaffected by these factors.

At the time, cigarette smoking was thought an unlikely culprit—indeed, with shades of "appeal to nature," it was actively seen as natural and thus a healthy habit. It was also a vice that had exploded in popularity with the advent of cheaper, more potent, mass-produced cigarettes, allowing deeper inhalation than previous variants like pipe smoking. However, the veneer of cigarettes as a healthy option slowly faded away as the evidence against them began to mount.

In 1929, Dr. Fritz Lickint unveiled convincing statistical evidence linking smoking with lung cancer; by 1939, Lickint's investigations culminated in a mammoth 1939 scholarly work, yielding a 1,200-page epic described by historian Robert Proctor as "the most comprehensive scholarly indictment of tobacco ever published," showing far beyond a burden of caution that smoking was not only linked to lung cancer but also implicated strongly in a range of other malignancies.

In tandem with Lickint's analysis, several other lines of evidence began to converge; statistical correlation consistent with the hypothesis that cigarette smoking could lead to cancer rapidly emerged, supported by the discoveries of carcinogenic compounds in cigarette smoke and the finding that cigarettes would induce cancers in laboratory animals.

This triumvirate of evidence became a rolling snowball. By the early 1950s, health boards around the world began warning customers of the dangers of cigarettes. In retrospect, this should have closed the book on the issue, but it was frustratingly dismissed and downplayed, leading to a distinct sense of inertia. Charles Cameron, director of the American Cancer Society, lamented in 1956: "If the degree of association which has been established between cancer of the lung and smoking were shown to exist between cancer of the lung, and say, eating spinach, no one would raise a hand against the proscription of spinach from the national diet."

Faced with overwhelming evidence of the toxicity of their wares and an increasingly negative public perception of smoking, the tobacco industry decided to dismiss science, armed with nothing save sound and fury. An internal memo that circulated around tobacco companies in 1969 attests to the cynical mindset and tactics in which they engaged:

> We are restricted in terms of ability to sell—in colleges and in vending machines. Our products are branded with a warning label. Our ability to advertise has been attacked on all fronts and has consistently deteriorated.... Doubt is our product since it is the best means of competing with the "body of fact" that exists in the mind of the general public. It is also the means of establishing a controversy.

To this end they were undeniably effective—doubt was indeed their product. When the public mood turned against them in the mid-1950s, the leading cigarette manufacturers turned to public-relations gurus to sow the requisite seeds of doubt. On January 4, 1954, an advertisement ran in over 400 newspapers across America. From its opening line to its flimsy conclusion, the now-infamous "Frank Statement" is pure rhetorical trickery, a master class in sheer verbal sleight of hand:

Recent reports on experiments with mice have given wide publicity to a theory that cigarette smoking is in some way linked with lung cancer in human beings. Distinguished authorities point out:

1. That medical research of recent years indicated many possible causes of lung cancer.
2. That there is no agreement among the authorities regarding what the cause is.
3. That there is no proof that cigarette smoking is one of the causes.
4. That statistics purporting to link cigarette smoking with the disease could apply with equal force to any one of many other aspects of modern life. Indeed the validity of the statistics themselves is questioned by numerous scientists.

It's worth analyzing these tactics in some detail as they are far from confined to history. Despite an outwardly reasonable opening, careful wordplay conceals layers of dishonesty. This is a classic strawman argument, conceding that studies in mice might perhaps bear some loose relation to human lung cancer. It was entirely deceitful; epidemiological and lab experiments had clearly demonstrated overwhelming evidence of a causative link between cigarette smoking and lung cancer, so invoking studies on mice was textbook misdirection, crafted to distract from the disconcerting human evidence. The masquerade of concern is even more insulting with the benefit of hindsight, when it transpires that cigarette companies' own internal research had shown their carcinogenic potential well before medical science began investigations.

The numbered points the statement attempted to ram home read like a checklist of rhetorical trickery. As we've already covered logical and informal fallacies in some depth, we won't dedicate too much time to skewering the first three items on the list (four if one counts the appeal to authority in the preamble). The first is a beautiful example of a *red herring argument*, or one that deflects from the issue. It does not matter whether other things may cause lung cancer; the issue at stake is whether smoking is a lung-cancer risk or not, and to what extent. This is merely a shameless attempt to deflect, dilute, and dissipate the issue. Point two was an outright falsehood as medical consensus had already converged to a verdict of guilty on cigarettes, as had the research of the cigarette companies themselves. Similarly, point three is another assertion that, even in 1954, was completely at odds with the evidence. Note also the shift from the singular "cause" in point two to the plural "causes" in point three—a change in the distribution of the premises, which would render the argument invalid even if the two points had themselves been true.

The final point was also a falsehood. The reality was that the first hints of carcinogenicity had arisen from careful statistical analysis. Lickint's painstaking work (and that of others) had ruled out confounding variables. The Frank Statement was a masquerade of concern, a cynical attempt to sow the seeds of doubt and neutralize the overwhelming scientific consensus. The aim was to create a perception in the public mind that the cases for and against smoking were equally valid positions. This was in effect an attempt to weaponize false balance; were there any doubt about this, the revelation of the odious "doubt memo" eventually laid this contemptible game plan bare.

This exploitation of false balance is the archetypal example of a *manufactured controversy*, a contrived disagreement to create uncertainty over issues where there is no real scientific dispute. For the cigarette companies, this tactic was effective, creating enough public doubt for acceptance of smoking to persist for decades, costing millions of lives. Nor were they the only ones to deploy such underhanded tactics; when evidence emerged that linked Reye's syndrome (a potentially deadly liver inflammation) in children to aspirin usage, manufacturers were able to delay mandatory warning for almost two years by using the same tactic. Similarly, despite ultraviolet radiation being a well-known carcinogen, organizations such as the now-defunct Indoor Tanning Association tried hard to downplay the scientific consensus.

It is worth acknowledging the dubious power of a well-aimed public-relations campaign. The Frank Statement was the brainchild of global PR firm Hill+Knowlton, which has quite an interesting track record in spin. In October 1990, during the run-up to the Gulf war, a tearful Kuwaiti citizen known only as Nayirah testified that she had seen Iraqi soldiers tear children from incubators and leave them to die. Such horrifying testament dominated news headlines and was mentioned several times by then President George H. W. Bush as emotive justification for war. It wasn't until after the war's conclusion in 1992 that journalist John MacArthur revealed "Nurse Nayirah" to be the daughter of Saud Al-Sabah, ambassador to the United States. Her emotive testimony was false, organized by Hill+Knowlton, whose (exceptionally cynical yet undoubtedly accurate) internal studies had shown that rhetoric about atrocities were most likely to influence American public opinion.*

Of course, we might be forgiven for thinking such tactics are a relic of another era and that, since we are a more sophisticated, perceptive public, such barefaced denialism couldn't possibly work on us. However, even now when the science is incontrovertible, sowing the seeds of doubt remains the refuge of the scoundrel; no matter how firm the evidence, parties with vested interests can undermine it by using the same cynical modus operandi that tobacco companies used decades ago. The motivation for this might be financial, as with tobacco companies. But ideological

---

* Not content with manipulating public opinion over a war that cost thousands of innocent lives, Hill+Knowlton has also lent its rhetorical skills to such admirable goals as representing the Church of Scientology and helping nations with human-rights abuse issues improve their reputation. It also orchestrated damage control for Theranos, which we encountered previously. Of course, Hills+Knowlton is just one PR company among many others that have mastered the exploitation of numerous fallacies covered earlier to lodge in the public psyche a message at odds with reality; this might be laughable were they not so incredibly effective at manipulating us.

convictions, whether political or religious, can be even more potent forces. There is, of course, nothing wrong with holding religious or political beliefs or with arguing for these ideals. But employing mendacious tactics rather than honest argument to promote one's viewpoint is completely unacceptable. False balance frequently lends itself to precisely this—and the machinations of the intelligent design movement are a fine example.

To many faiths, the intricate nature of life on Earth shows the hallmarks of design, usually the handiwork of a god. This teleological argument, or argument from design, has a long pedigree; the earliest recorded version was discussed by Socrates, but even then it was ancient. By the medieval era, it was a staple of Christian belief, with a version of the argument appearing in the Bible through the words of Paul the Apostle. The thirteenth-century Christian philosopher Thomas Aquinas even presented it as one of his five logical arguments for God's existence. To theological minds of the Middle Ages, this was the only explanation that could account for the sheer scope of life on the planet and the beautiful intricacy with which it functioned. In the abundance and elegance of life on Earth, they saw the fingerprints of an ethereal master craftsman.

Despite superficial appeal, though, the logic is rather flawed and has been unpicked by Scottish philosopher David Hume and others over the centuries. The teleological argument's last trump card was the seemingly inexplicable variety in the natural world, surely too complex to have emerged unscripted. But by the twilight of the nineteenth century, Darwin's evolutionary theory removed the need for some external agent sculpting creation, demonstrating instead that the crucible of the environment itself shaped how species differentiated over time; the complexity of all life on Earth had no need of an artisan deity.

Evolutionary theory itself is silent on the question of God, but to biblical literalists it remained an affront. Early twentieth-century America saw a growing chasm between modernist and fundamentalist factions of the Presbyterian Church. The teaching of evolution became a flashpoint, with the emerging fundamentalist faction viewing it as rank apostasy. To William Jennings Bryan, it was a personal vendetta. Formerly a state representative, he had thrice run unsuccessfully for president as the Democratic Party's nominee. After the third defeat, he returned to his faith, becoming convinced evolution was an insult to God. Bryan's views were sufficiently fringe to be rejected by his peers—but, although he couldn't sway his own church, he retained enough political clout to lobby states directly. A handful of states duly banned the teaching of evolution for decades, with Tennessee only repealing this ban in 1967.

The end of educational censorship should have been enough to see the edifice of creationism crumble. And so it might have been if, in the mid-1980s, law professor Phillip E. Johnson had not undergone a dramatic religious conversion following a divorce, emerging from the experience as a leading light of the Intelligent Design (ID) movement. The "Intelligent" part of ID is distinctly oxymoronic—these ideas were simply creationism posturing in scientific garb. Still, the movement grew, strengthened by the formation in the 1990s of the Discovery Institute, dedicated to "a science consonant with Christian and theistic convictions."

To achieve this, they crafted the "wedge strategy," designed to wedge their beliefs into public discourse. The central idea was to spin evolution as merely a theory, insisting their ideas were equally valid and should also be taught in schools. To the unsuspecting, this might seem reasonable; evolution is indeed "just" a theory, so why does it deserve special status? But the word "theory" is rife with ambiguity, dependent on context. In colloquial English, it implies supposition or speculation. But in scientific parlance, a theory is a thoroughly tested hypothesis that best explains all the observed data, with both explanatory and predictive power. A scientific theory is not just conjecture but something that has been vigorously tested and is supported by multiple strands of evidence. Evolution is "just" a theory, in the same way the germ theory of disease or the theory of relativity are "just" theories: widely accepted scientific explanations for phenomena underpinned by overwhelming evidence.

Rather disingenuously, the Discovery Institute conflated these two different definitions, hoping false equivalence would bolster their religious philosophy. This "Teach the Controversy" campaign was slammed by the American Association for the Advancement of Science as a manufactured controversy, for erroneously claiming that evolution was scientifically controversial. Despite the leaking of the wedge strategy online in 1999, the Dover Area School District of York County, Pennsylvania, began teaching ID alongside evolution, until a 2005 legal challenge ruled that teaching it as on a par with scientific theory was absolutely unjustified.

Outside America, creationist posturing might be laughable, but the ongoing "debate" over climate change is the epitome of weaponized doubt. Evidence that human activity is driving radical climate change is simply overwhelming. Since the dawn of industrialization, the millions of tons of CO_2 we have pumped into the atmosphere have steadily and worryingly drawn average global temperature persistently upward. Despite the clear scientific consensus, denialists still try to pour doubt on the issue in order to disguise the fact that their position is untenable. Climate change denial is not a fringe belief—it is passionately held by a large swath of the population. Part of the reason for this is false balance. While the scientific evidence for climate change is virtually incontrovertible, coverage rarely captures this reality. Boyce Rensberger, director of the Knight Center for Science Journalism at MIT, noted that "balanced coverage of science does not mean giving equal weight to both sides of an argument. It means apportioning weight according to the balance of evidence."

Sadly, false balance on climate science has long been the default position: a survey of 636 articles from four US newspapers published between 1988 and 2002 indicated that most articles had devoted equal coverage to a small collection of climate change denialists and those who represented the scientific consensus. This situation is not specific to America either. Even British state broadcaster BBC–renowned for its impeccable science programming–is not immune to this mistake. A 2011 trust report harshly criticized the BBC's "undue attention to marginal opinion" on the subject of man-made climate change. The same report found that, despite the overwhelming scientific evidence that human activity is driving climate change, several BBC shows fell victim to an "over-rigid application of editorial guidelines on impartiality" with the net result being far too much airtime for climate change denialists. A follow-up report published in 2014 found that this key conclusion "still resonates today."

The net consequence of this is that climate science is affected by a stark consensus gap: While the scientific consensus on climate change is virtually unanimous, the public is left with the impression that climate science is somehow controversial. In a 2013 study, members of the public estimated that just over half of scientists agreed on climate change; the real figure is nearer 100 percent. This lingering doubt suppresses action on climate change, but in recent years there have been indications that things are slowly improving. A 2017 paper found that, while "media coverage across countries and media outlets moves closer to the basic scientific consensus on climate change," there was still undue input from denialists, and niches where climate change denial was rampant. Typically, these niches were found to be columnists in conservative outlets–an unsurprising finding given our prior discussion of ideologically motivated reasoning in climate change denial.

Of course, balanced coverage is welcome–and journalistic impartiality laudable–but balance must be proportional to the strength of evidence for each position. It is foolhardy in the extreme to naively treat ideas with vastly disparate evidence bases as equally valid. All this does is elevate dreadful ideas at the expense of good ones, leaving us open to manipulation by devious elements. While it might seem tempting to berate media outlets as systematically inept, this would be an unfair and misguided dismissal, for without them we are at the mercy of questionable sources.

Traditional outlets thus have a critical role to play in conveying accurate information and viewpoints. Their ability to push a standard for fact-checking and quality control is lacking in more fragmented modern media, yet it is more vital now than ever. Impartiality is a bulwark against the undue sway of our increasingly partisan sources. But engaging in false balance undermines this strength and risks giving debunked, dangerous fringes an air of legitimacy and the oxygen of publicity. Ultimately, such sophism leaves us all more divided and less informed.

16

TALES FROM THE ECHO CHAMBER

*AIDS Denialism, the Filter Bubble, and the
Polarizing Perils of the Echo Chamber Effect*

Predicting the future is no easy undertaking. The 1936 American presidential election, which pitted incumbent Franklin D. Roosevelt against Kansas governor Alf Landon, is a case in point. The Great Depression had lingered for eight long years and the country remained in a fragile state. Roosevelt had won the presidency four years earlier with his "New Deal" economic policies. Some of his measures had proved popular with the electorate, such as social security and unemployment benefits. Even so, he was struggling to push his reforms through Congress and the courts. Some opposed him on fiscal grounds; Landon accused Roosevelt of being hostile to business. Political pundits predicted a close race between the two men.

Chief among the publications vying to predict the new president was the *Literary Digest*, a general-interest weekly that had correctly predicted the result of each election since 1920. The *Digest* was eager to maintain its unbroken record of success, but a changeable public mood rendered this far from a simple undertaking. They quite reasonably decided that the only way to predict the future was to enlist an unprecedentedly massive sample group. With admirable determination, they polled a staggering 10 million people—roughly a quarter of the American electorate at the time. In the August issue, the editorial confidently asserted that their poll would yield the result of November's forthcoming election "to within a fraction of 1 percent the actual popular vote of 40 million." A total of 2.4 million people responded to the poll, and from these results the *Digest* confidently predicted the seemingly inevitable result: Landon would walk away with the election, commanding a comfortable 57 percent of the vote to Roosevelt's 43 percent.

But history does not agree with this projection. Landon did not become the 33rd American president; nor was it even a close call. Instead, Roosevelt was reelected in a landslide victory, capturing every state except Vermont and Maine. Even more curious was that a young statistician named George Gallup had predicted exactly this, without the machinery of the press at his disposal. Gallup had made these predictions with a relatively paltry 50,000 people to survey, about one-fiftieth of his rival's sample group. On the face of it, this might fly in the face of all we know—we've already seen that a larger sample size begets more accurate results. So how then did the *Literary Digest* get it so spectacularly wrong when armed with such an impressive number of respondents?

The root cause of the discrepancy was a subtle but crucial one. To solicit the greatest volume of data, the *Digest* had sought responses from three readily available lists: its own readership; the telephone directory; and the automobile registry. But therein lay the problem. If you belonged to one of these groups, the chances are you were far wealthier than the average American of the time. This was compounded by the fact that only a self-selected proportion of those polled chose to respond. This rendered the enormous sample group irreparably skewed, and ultimately not representative of the true electorate. Adding insult to injury for the beleaguered paper, Gallup predicted this confounding influence and, as well as the correct outcome of the election, he was even able to predict the results of the *Digest*'s poll, based upon their sources and their likely bias. The faux pas by the *Digest* was a costly one, paving the way for it to fold, shrouded in ignominy. Gallup, by contrast, went on to found an eponymous polling company that is still going strong to this day.

The crucial lesson here has far more urgency today than it had in the 1930s. The perceptions of the *Digest* staff were distorted because, even with their colossal sample group numbering in the millions, they had failed to acknowledge that their audience was not representative of the whole. Of course, this mistake by the *Literary Digest* entails aspects of an entire array of related problems, some of which we've touched on already: confirmation bias, skewed samples, and even perhaps a tangible amount of wishful thinking. But in the broadest possible terms, the greatest error lay in assuming that the voices in their community were representative of the general scenario. And in this respect, the magazine's collapse is a cautionary tale. It is all too easy to reside in an echo chamber of ideas similar to our own and, whether by accident or design, render ourselves deaf to facts that might undermine our comforting perceptions, sometimes leading to tragic consequences.

Echo chambers have always existed to some extent, be they newspapers with political agendas or biases in television shows. But today, the problem is more acute than ever before, and potentially even more damaging. The somewhat counterintuitive reason for this is the extreme level with which we have embraced the internet. This might seem paradoxical—naively we might think that the internet allows complete freedom of expression and would expose us to a plethora of voices that traditionally we may never have encountered. Certainly that was the heady optimism with which we embraced this emergent technology. But the reality is somewhat murkier. We live in an age of algorithmic filtering and directed advertisement, and this in turn directly shapes what information is tailored for us. Social media sites may be free to use, but they tend to rely on advertising revenue and, accordingly, it is in their interest to direct this as best as possible. After all, we are the product.

This is an inherently sycophantic operation, seeking to flatter us with content and opinions we are likely to view favorably. These algorithms curate precisely what we want to see and predict what we are likely to find agreeable while simultaneously exiling challenging information, views, and ideas beyond the confines of these comforting bubbles. They are fine-tuned to deliver us more of what we want and less of what we don't. And this gatekeeping has a tangible impact on how we receive and process information, especially when recent data suggests that social media has become one of our primary (and, in some cases, only) news source. Internet activist Eli Pariser defined this as the "filter bubble," criticizing social media outlets for fawning to narrow self-interest by offering "too much candy, and not enough carrots." With 1.6 billion active users, Facebook provides an ample case study in how self-selection occurs. It is the world's most popular social media site, with a business model reliant on its ability to direct advertisements to a vast user base. It is also notorious for employing user data to shape the experience.

The internet is still in its infancy, and the impact of all this is still contentious. Given the ambiguity of the data, I'd be remiss to decry filtering as a great social evil. Technology is never intrinsically good or bad; what matters is how it is employed for different applications. There are instances where the propensity to preempt a user's interest is desirable. A metallurgist and a hard rock fan, for example, would get markedly different results if they both searched "types of metal" and this stratification* would save them both time. The problem arises when we use social media as a barometer for our ideas and opinions. If only supporting information is selected and challenging views are discarded, then we may mistakenly construe support of a position online with real-world support, even if our idea is extremely flawed. This has the potential to induce a communal form of confirmation bias, with all the apparent support being little more than argument from anecdote. These are, as we have seen, inherently shaky grounds for decision-making and prone to lead us to poor decisions.

The crude nature of these filters can lead to bizarre stances. Facebook remains the most popular social media site on the internet but, in striving to rid the platform of nudity, clumsy implementation often undermines it entirely. One ignoble example is the company's response to an article by Norwegian writer Tom Egeland's 2016 article on photographs of war that had altered public opinion. It shortlisted the harrowing image "The Terror of War" by photographer Nick Ut. This famous shot focuses on the haunted

---

* The pun wasn't intended, but I'm opting to leave it in.

figure of nine-year-old Kim Phúc, naked and terrified after a napalm attack at the height of the Vietnam War. The horror of the image implanted it into our cultural consciousness, and it won the 1973 Pulitzer Prize for Photography. Rigidly applying their guidelines without any respect to the context, Facebook banned Egeland despite the historic significance of the photo. A stern rebuke from the press and politicians in Norway and the wider world eventually forced Facebook to climb down. Yet mindless filtering remains a huge problem for the platform, frequently leading to the banning of pages on topics such as breastfeeding, medicine, and art, while hate groups' pages often remain untouched.

Blame for the wider trend toward self-selection cannot be laid solely at the feet of internet giants, though; it transcends social media, tapping deep into our human desire to establish our own selective realities. Despite fears over algorithms deciding everything, a study of 10.1 million Facebook profiles found that users' explicit specifications for what they wanted to see had a greater impact on their feeds than the standard sorting algorithms and was a greater factor in limiting opposing views. In essence, we do most of our selective pruning ourselves. Indeed, the online echo chamber mightn't even be that much greater than the real-world one we inhabit all the time—just more prone to extremism. A 2016 study in *Proceedings of the National Academy of Sciences* (PNAS) found that misinformation thrived online particularly because users tended to "aggregate in communities of interest, which causes reinforcement and fosters confirmation bias, segregation, and polarization."

In some respects, it's not surprising that internet users cluster into self-affirming echo chambers. That this might occur was predicted as early as 1996 by MIT researchers Marshall Van Alstyne and Erik Brynjolfsson. They dubbed it *cyberbalkanization*, a play on the fractured state of the Balkans. In their paper, the authors acknowledged the potential for the nascent internet to transcend traditional divisions and boundaries but also warned of a potentially detrimental insularity:

With the customized access and search capabilities of IT, individuals can focus their attention on career interests, music and entertainment that already match their defined profiles, or they can read only news and analysis that align with their preferences. Individuals empowered to screen out material that does not conform to their existing preferences may form virtual cliques, insulate themselves from opposing points of view, and reinforce their biases. Internet users can seek out interactions with like-minded individuals who have similar values, and thus become less likely to trust important decisions to people whose values differ from their own. This voluntary balkanization and the loss of shared experiences and values may be harmful to the structure of democratic societies as well as decentralized organizations.

This proved a remarkably percipient insight, coming years before mass internet use and the advent of social media. The net result is a vicious circle, where the positive feedback derived from self-selection skews perception, deviating from reality even further. Take Twitter as an example of this. As a microblogging platform, posts are by default set to be public. However, this does precious little to offset the echo chamber effect, as users populate their news feed with people whom they've chosen to follow.

The reality is that media channels of distribution have changed almost beyond recognition in a small window of time. In the past, information emerged from newsrooms the world over, with a team of journalists and editors working to maintain standards and to check the facts of any story. This didn't always succeed and there remains ample scope for improvement. Yet, for all the flaws of the conventional model, it provided some modicum of quality and standard of journalistic integrity. In traditional media, stories, opinions, and features were clearly demarcated and typically provided in context. But in the few short years since the turn of the millennium, this has all changed dramatically.

Most of us now get our news online. And in general, we're not getting it from the digital hubs of traditional media outlets but diffracted through the prism of social media. The upshot of this is that we have become our own curators with a tendency to cherry-pick items utterly devoid of context. And when we become our own editors, we tend to reinforce our inherent biases rather than challenge them. In curating our own news, we tend to gravitate to that which conforms to our perceptions and assures us that our view of the world is correct. With the abundance of sources available, we can easily pick a handful of articles extolling a belief we already have, supplementing this with a plethora of blog posts or YouTube videos. Rather than simply receive information in context, we effectively compile a digital scrapbook of that which we would like to be true rather than that which actually is true.

Stripped of context and barriers, the most extreme voices at either end of the spectrum dominate, staking out mutually opposed camps. Any issue seen through this reductive lens is rendered in binary: either "good" or "bad," "right" or "wrong." Of course, as we've seen before, this is often a false dilemma, ignoring the spectrum of views one might hold. But the fact that one person can simply run only with the sources that agree with their views while another can do precisely the same thing in reverse stamps a deeply subjective tint on each argument, making it harder than ever to establish any form of commonality. Worse, this can be so pronounced that it can become impossible even to establish agreement on basic facts, resulting in ever-increasing polarization. The problem is laid bare by a 2016 report from the Columbia Journalism School on social media's impact:

> Qualified news is mixed with unchecked information and opinions. Rumors and gossip get in the flow. We call this digital fragmentation. Journalistic companies . . . are forced to cut costs, lowering their capacity to offer more corroborated news, context, and analysis. One effect of digital fragmentation is polarization.

Non-fact-based opinions and rumors accelerate the behavior of
quickly taking a short cut to "like" or "dislike." . . . People may be
losing the skills to differentiate information from opinion.

A Pew Research survey found that 66 percent of millennials (defined
as those born from the mid-1980s to the present) derive their news
primarily from social media. More tellingly, similar research indicates that around 40 percent of users have purged a social media
contact for having conflicting political opinions—suggesting that,
not only do we love our walled gardens, but we prune them to be
ever more homogeneous. This blatant confirmation bias blinds us
to the often-pertinent objections and viewpoints of others, and lulls
us into an ever more damaging orthodoxy. This problem transcends
social media, afflicting even search engines such as Google, which
deploy a variety of algorithms to rank pages, with popularity also
playing a massive part in ordering search results. But as search
engines build up a huge repository of user data based on what their
previous searches have suggested, it risks catering to our whims so
much that it can exacerbate the problem further.

Echo chambers can threaten our physical health too, and the
alternative medicine community is one such galling example.
We've previously mentioned whatstheharm.net, which records
some of the heartbreaking ways people can succumb to the lure of
alternative medicine in lieu of effective treatment. An echo chamber of confirmation bias can encourage people to believe the most
outlandish things. To see how dangerous this can be, we need only
look to an especially tragic form of groupthink: AIDS denial. The
story of how the human immunodeficiency virus (HIV) emerged
from the jungles of Africa to wreak deadly havoc across the world
is an alarming part of history. The origins of HIV are still not
entirely clear, but we now know it is closely related to a similar
immune deficiency that affects nonhuman primates in Central and
West Africa. Somewhere between the late nineteenth and early
twentieth centuries, this virus crossed the species barrier, afflicting

hunters who sought bush meat for protein, with biting by chimps and blood-to-blood contact during butchering possibly facilitating transmission.

Once infection is established, the carrier initially shows no symptoms. Slowly, increasing viral load begins interfering with the immune system. Over a number of years, victims become progressively less able to fend off the myriad immunological threats that humans encounter each day. In time, this leads to Acquired Immune Deficiency Syndrome (AIDS) where the patient's CD4+ T-cell count falls so critically low that they're unable to fend off opportunistic infections and even cancers, affecting nearly every organ system. Without treatment, this is most often a death sentence. For decades, the virus had been confined to deep jungle on the African continent. But with its long incubation, it was perhaps inevitable that it could not be contained forever. Early cases occurred in isolation as early as 1959 in what was then called the Belgian Congo, and in an American teenager who died in 1969. The most common strain, HIV-1, arrived in Haiti from the Democratic Republic of Congo sometime in the mid-1960s. Around 1969, a single infected individual traveled from Haiti to the US. Unbeknownst to anyone at the time, that voyage would prove to be an extraordinarily fateful event, for that single patient became the genesis of almost all HIV cases outside of sub-Saharan Africa.

The initial latency of the virus aided its sinister perpetuation, until slowly it took root. In June 1981, the Centers for Disease Control (CDC) was perplexed by a cluster of five unusual cases in Los Angeles, where young and otherwise healthy men fell ill with a form of pneumonia usually only encountered in the extremely immune compromised. Over the next few months, young gay men began dying of opportunistic infections related to a depressed immune system. Kaposi's sarcoma, a once-rare form of cancer, suddenly became common, afflicting men across the US with its distinctive red nodules. An avalanche of new cases ensued, and by 1982 it had been dubbed GRID—gay-related immune deficiency. But throughout the homosexual community, it had a more evocative name: gay cancer.

Within months, this moniker became obsolete. Cases began to emerge in cohorts of intravenous drug users, hemophiliacs, and Haitians. By 1983, the crisis had spread around the world and the CDC had a new name for it: AIDS. In January of that year, Françoise Barré-Sinoussi and Luc Montagnier at the Pasteur Institute in Paris presented evidence of a T cell–killing retrovirus isolated from the lymph system of an AIDS patient, a finding for which they eventually shared a Nobel Prize. This was independently discovered by Robert Gallo, an American biomedical researcher who showed that this virus could lead to AIDS. In time, it received its modern name: human immunodeficiency virus—HIV.

With the cause identified, the hunt for a cure began in earnest. The scale of the pandemic was laid bare on January 14, 1986, when Anthony S. Fauci of the National Institute of Allergy and Infectious Diseases gravely informed America that over a million people in the US had been infected, with that number expected to double or even triple within the decade.* Scientific teams worked at breakneck speed to counter the virus, and by 1987 the first antiretroviral treatment for HIV, AZT, became available.

The advent of antiretroviral treatments (ART) marked a watershed moment in the fight against the illness. In the early days, some researchers questioned whether there were enough grounds to confidently assert that HIV led to AIDS, but the evidence for this came quickly to the point where such a conclusion was incontrovertible. One notable exception to this consensus was scientist Peter Duesberg, a then-eminent researcher and member of the prestigious National Academy of Sciences (NAS). Duesberg instead claimed that HIV was harmless and that AIDS was in fact a proxy for some homosexuals to use drugs such as poppers (alkyl nitrites).

---

* Fauci has recently become a household name worldwide for his work in 2020 on the White House Coronavirus Task Force, portrayed by Brad Pitt on *Saturday Night Live,* and lambasted by conspiracy theorists. But this goes to show his expertise long predates and exceeds his current fame.

More alarmingly, and without any evidence, Duesberg claimed that AZT itself *caused* AIDS, despite the growing mountain of evidence that showed his contention to be groundless. As a member of NAS, Duesberg invoked his right to have an article published without peer review in order to advance his increasingly ill-supported views. While this was an abuse of this right, the journal editor wearily conceded to his demand, stating: "If you wish to make these unsupported, vague, and prejudicial statements in print, so be it. But I cannot see how this would be convincing to any scientifically trained reader."

Lacking supporting data, Duesberg relied on his reputation and loaded oration. Other researchers easily took his proclamations apart with surgical precision, making the once-respected scientist a pariah. John Maddox, editor of *Nature*, dismissed Duesberg's continued demands to advance his position without evidence, declaring he had "forfeited the right to expect answers by his rhetorical technique." But the seeds of doubt planted by Duesberg and his ilk gained currency in alternative circles; certainly his suggestion that AZT causes AIDS resonated with those who believed that nefarious big pharmaceutical companies were involved in some odious ploy to make money—a distrust amplified by a rumor that AIDS had been manufactured to kill undesirables.

Such narratives gained traction in the hardest-hit communities, and remain hard to shift. Even today, up to half of Black Americans believe HIV was a man-made weapon, designed to curb the poor, Black, gay, and Latinx populations. These perceptions must be seen in context; Black Americans have long been victims of systemic inequality and accounted for a disproportionate number of new HIV infections. In communities ravaged by the disease, these beliefs sometimes became widely held. The major downside to these views is not merely some abstract epistemological concern; they can act as barriers to treatment and prevention, as believers feel they will be made victims no matter what measures they take and so may be less inclined to take preventive measures.

The male homosexual community was hit hardest by the disease and the stigma that went with it. In the wake of the sexual revolution, the gay community tended to experiment more, averaging 11 partners a year. Common sex practices also came with a higher risk of disease transmission, due to the susceptibility of anal tissue to tearing during sex. When panic over AIDS reached its zenith, gay men were debased and discriminated against. Even as scientists labored admirably to find a control, Ronald Reagan's White House largely ignored the crisis, with pleas by the CDC for funding routinely denied. By the time Reagan acknowledged the extent of the crisis in 1987, almost 21,000 people were dead. By the time he left office in 1989, this figure had risen to over 70,000. This apathy by the White House was perhaps telling of a more toxic mindset: a callous disregard for the worst-afflicted. On top of this, a mixture of widespread ignorance and outright homophobia meant that gay men were frequently depicted as unclean. Religious zealots even proclaimed AIDS as some kind of karmic retribution for defying their perception of God's will, with popular Baptist minister and television personality Jerry Falwell proclaiming that "AIDS is not just God's punishment for homosexuals, it is God's punishment for the society that tolerates homosexuals."

Faced with such bigoted hostility and feeling a keener sense of loss than any other subculture, it's not surprising that some gay men rejected the scientific consensus. Instead, a minuscule but vocal minority borrowed liberally from ideas by Duesberg. In 1992, London-based activist Jody Wells created *Continuum*, a magazine that took a position of AIDS denialism, claiming that HIV did not cause AIDS. Wells believed that AIDS was a conspiracy rooted in homophobia. *Continuum* went so far as to question whether HIV really existed. The magazine's staff refused to accept the efficacy of HIV drugs, and promoted alternative cures in their stead. *Continuum* finally ceased publication in 2001 after its entire editorial staff succumbed to AIDS-related illnesses.

Nor were *Continuum*'s staff the last casualties of this belief; many other prominent AIDS denialists needlessly died in subsequent years. But the deleterious consequences of AIDS denialism persist even to the present day. Thabo Mbeki became president of South Africa in 1999, adopting a position deeply sympathetic to AIDS denialists, despite the high HIV infection rate in the country. During his stint in office, he roundly ignored scientific advice, and instead drew heavily upon the beliefs of the denialist movement. He even denied antiretroviral drugs to HIV-positive patients, decrying them as "poisons." In keeping with the logic of an echo chamber, Mbeki appointed a health minister with similarly fringe beliefs, Manto Tshabalala-Msimang. She advocated garlic, beetroot, and lemon juice rather than ART to treat AIDS, a bizarre course of action that led deeply frustrated members of the medical community and South Africans to bestow upon her the barbed title of "Dr. Beetroot." In response to the Mbeki government's dangerous policies, 5,000 scientists and physicians signed the Durban Declaration in 2000, stating unambiguously that HIV causes AIDS and criticizing the reliance of the South African cabinet on the denialist fringe.

Such concerns fell on deaf ears. Mbeki instead commissioned a scientific panel consisting overwhelmingly of AIDS denialists, including Duesberg. The scientific consensus was again ignored, with the panel advocating holistic and alternative medicine instead of ART for HIV. The stubborn insistence of Mbeki's government on listening only to those with similar fringe views rather than heeding medical advice came at a staggering cost: By the time he left office in 2008, Mbeki left an appalling legacy of between 343,000 and 354,000 preventable AIDS deaths.

This is the essence of the worst danger of the echo chamber: Group consensus cannot circumvent reality. The echo chamber is simply cherry-picking and conformation bias writ large, but perhaps most important, it is something we must strive to be constantly aware of in our own social circles. Received wisdom should not be exempt from critical examination; indeed, if the wisdom is robust, it should withstand it. Echo chambers have always been a problem, but in the internet era they threaten to get worse, increasing fragmentation and the ensuing polarization of the modern world. We live in an era where, no matter how outrageous, archaic, or dangerous a person's beliefs might be, that person can with ease find a community that echoes their worldview precisely.

That groups have similar beliefs may generate some feeling of social cohesion, but all this is for naught if it renders members deaf to the intrusions of conflicting evidence. Of course, information doesn't exist in a vacuum; to really appreciate the challenges we face, we have to explore how information itself is disseminated—and how it can be readily skewed.

THE OUTRAGE MACHINE

The Sandy Hook Atrocity, Soviet Interference,
and the Art of Dezinformatsiya

For residents of Newtown, Connecticut, December 14, 2012, is an inauspicious and painful date. That morning, 20-year-old Adam Lanza crossed the threshold of Sandy Hook Elementary School. Unbeknownst to anyone, Lanza had just executed his own mother with a bolt-action rifle. Nor could anyone have known that, immediately after this act of matricide, he had augmented his armory with his mother's Bushmaster rifle. Using this pilfered weapon, Lanza shot his way through the doors of the school. By the time Lanza pulled the trigger with the muzzle against his own head, he had already murdered 20 children ages six and seven, and six staff members. It took less than five minutes for the quiet village of Newtown to be changed irrevocably, becoming the epicenter of one of the most atrocious school shootings in US history.

There was no rhyme nor reason, no higher purpose to this crime, no master plan—just the actions of a deeply disturbed young man, leaving a shaken nation and devastated families in his wake. Before the bodies of the victims were even cold, the cynically self-serving National Rifle Association (NRA) was in damage-control mode, arguing against all evidence that ease of access to such weaponry had nothing to do with the tragic events that ensued.* In an instant, Sandy Hook became a byword for the long-standing debate over gun-control laws, and the acrimonious row began anew. Such a turn of events was hardly unexpected; America has long had a polarized stance on guns, and each fresh horror brings the conflict

---

* The NRA is almost certainly incorrect in its assertion. Evidence to date suggests strongly that access to firearms does indeed dramatically increase the risk of their being used for nefarious purposes and makes people collectively less safe. This is not news to most of the world, where access to firearms is restricted, but is somehow still contentious in America.

into focus again. No one doubted that the massacre would provoke passion. But no reasonable human being could have anticipated which party would incur the true wrath of a vocal underbelly—the bereaved parents of the victims.

News of the slaughter had no sooner hit prime-time coverage than the cogs of the conspiracy machine began their inextricable grinding motion toward outlandish conclusions. The paranoid denizens of internet forums dedicated to conspiracy were, as usual, unanimous in their rejection of the official story. Instead, they forged an alternative version of the truth: that the Sandy Hook murders were a fabrication, an outright concoction staged to garner sympathy for gun-control legislation. The entire tragedy was, in their minds, a smokescreen, an elaborate ruse to sway a gullible public. To most of us, this notion is so unbelievably bizarre that we struggle to wrap our minds around the tortuous nature of reasoning required to get to that conclusion. Even so, the belief that Sandy Hook was a so-called "false flag operation" rapidly took root.

The conspiracy angle was born the moment rolling coverage began. As we've seen, eyewitness accounts often conflict and small details can be mangled in the frantic news cycle. Sandy Hook "truthers" latched on to each minor and mundane reporting error as proof positive of a cover-up. Any perceived inconsistency, no matter how trivial, was rapidly absorbed into truther canon. In a textbook display of confirmation bias, this eagle-eyed analysis was wholly skewed and the multitudinous glaring holes in its narrative were ignored in order to preserve belief. Within only two days of the shooting, the first YouTube video claiming the massacre was a fake had been uploaded. It was to prove a vanguard for the thousands that followed.

A whole community of similar-minded individuals began combing media accounts for proof that the entire affair was a hoax designed to strip them of their Second Amendment rights. This kind of shoddy logic was exemplified by figures such as Orly Taitz, who asked: "Was Adam Lanza drugged and hypnotized by his handlers to make him into a killing machine as an excuse as the regime is itching to take all means of self defense from the populace before the economic collapse?" As it so often does, the dubious imprint of media coverage and celebrity endorsement can project a bad idea far beyond the confines of its merit. Radio personality and conspiracy theorist Alex Jones played this ignoble role with gusto, expounding the threadbare logic to his devoted audience of more than 2 million listeners. A mere five days after the shooting, Jones's hugely popular website InfoWars.com asked whether the massacre was "all part of an evil pre-conditioning program."*

These accusations, shrouded with a fig leaf of inquiry, were only the beginning. Soon a slew of variations on the theme emerged malformed on the internet. But more alarming than the mere delusion was the vitriol its twisted internal logic inspired. For the truthers' version of events to be correct, the families themselves had to be the worst kinds of fraud. In this telling of events, the inconsolable parents on the evening news were transformed from people deserving the utmost sympathy to mere actors, duplicitous agents of a nefarious government. To those invested in the narrative, they were less than human and the worst kind of traitor. The truther movement christened these parents "crisis actors," a label that would prove a modern mark of Cain.

Attacks on the bereaved families came thick and fast. Sometimes they masqueraded as questions, albeit of a most inappropriate and callous nature, demanding graphic details of how the children died. More often, they comprised bald assertions and unhinged threats. Gene Rosen was one such target; during the attack, he sheltered six students and a bus driver in his home amid the hail of bullets. Despite his heroism, Rosen remains the victim of an active online harassment campaign by those who insist he is complicit in the government cover-up. Following an interview with CNN, Robbie

* This "Just Asking Questions" tactic of framing wild accusation or outlandish speculation as questions is such a common bad-faith technique that it has its own descriptive acronym: JAQ-ing off.

Parker—who had just lost his daughter Emilie in the attack—was accosted by the ominously named "Operation Terror," who had the single-minded audacity to declare his grief an act.

Nor were these altercations confined solely to internet outbursts. Some people went so far as to visit Newtown and harangue residents with accusations. Mere days after the shooting, an individual pretending to be a relative of Lanza arrived in Newton, uploading videos he claimed proved the entire town was a hoax. Others penned vicious letters to the families. Victoria Soto, the teacher who died heroically in the attack, had sacrificed herself to shield her students. After her death, her sister was confronted on the street by an individual who thrust a photo of her deceased sibling into her face, demanding she confess that the shooting had never taken place and admit that Victoria had never existed. In May 2014, a memorial was stolen from a playground dedicated to victims Grace McDonnell and Chase Kowalski. Not content with this violation, the thief called Grace's parents and proclaimed the deaths a hoax. He was eventually apprehended, and the memorial was returned to its rightful location.

Among the most egregious demonstrations of this unrelenting zealotry was the deplorable treatment of Lenny Pozner. Before that awful winter morning, Pozner's son Noah was a happy, photogenic child, frequently pictured with his adoring father. After his death, these pictures became a poignant reminder of the unfathomable human cost behind the grim statistics and a powerful image in the gun-control debate. The Sandy Hook truthers immediately attacked them as fake, stating they were designed to manipulate public emotions. Most of them continued unabated with the frankly ridiculous suggestion that Noah had never existed. Understandably perplexed by this incomprehensible denialism, Pozner released Noah's death certificate in a bid to reason with those dismissing the bereavements the Newtown families had suffered. Noble as this plea for a modicum of compassion and rationality was, it went entirely unheeded. True to form, the truthers decreed the certificate a hoax, responding with the most personal and spiteful abuse they could muster. Parents of victims were drowned under a deluge of emails, letters, and phone calls denouncing them violently, mocking their children's deaths or declaring them fictions. To break the spirits of families further, the truthers slurred them without any punishment. A typical YouTube entry is entitled "Fuck You, Lenny Pozner," tagged with the labels "Noah Pozner, Hoax, Pedophile." His mail haul, like many of the Newtown victims, is similarly graphic with a frenzied torrent of slurs and rabid accusations—and frequently, taunts and insults about Noah, an innocent child who had hurt no one before his life was violently taken.

Faced with this barrage, residents responded with admirable strength of character. Tired of their deceased children bearing the brunt of this unhinged abuse (often with racist or sexual overtones), Pozner and the other parents set up HONR—an organization with the mission to "bring awareness to the cruelty and criminality of Hoaxer activity and, if necessary, criminally and civilly prosecute those who wittingly and publicly defame, harass, and emotionally abuse the victims of high-profile tragedies and/or their family members. We intend to hold such abusers personally accountable for their actions, in whatever capacity the law allows." Like the similarly named 9/11 deniers, the "truther" epithet is a perversion of the word; these individuals claim to seek a hidden truth, but the reality is that they are only concerned with propagating a version of events that suits their worldview. HONR does not dignify these denialists with the "truth" label, instead choosing a more apt term: hoaxers.

This two-pronged ordeal the families of those slain at Sandy Hook have had to endure is resolutely unfair and, to anyone outside the bubble of conspiracy groups, deeply unhinged, but it is also a sobering exemplar of quite how dark the echo-chamber effect can become. People will always believe strange things in isolation, but it takes a certain critical mass before a group can feel confident enough in their beliefs to justify harassment. With Sandy Hook hoaxers, these fringe views found a crucible in the form of internet conspiracy subcultures. And in these bubbles, completely ludicrous views were not countered but encouraged. Without this essential criticism, members hear only their own beliefs parroted back to them—a self-sustaining reaction nurtured by a chorus of similar voices. To comprehend this, we need to understand the problematic issue of sources. The internet has put the entirety of the world's information at our very fingertips . But this same freedom also allows dubious information to propagate at dizzying speeds. And often, people lack the requisite critical skills to differentiate between reliable sources and those of dubious merit.

This is not a trivial problem and our propensity to find ideological echo chambers means that these problems feed into one another. What happened in the wake of Sandy Hook was a grim reminder of the dangers of twisted sources. Despite the ample media coverage of the incident, such sources were rejected outright by conspiracy theorists, who shun what they dismissively label "mainstream media." Instead, prominent conspiracy theorists like Orly Taitz and Clyde Lewis gave a more odious explanation. Alex Jones in particular was instrumental in perpetuating the narrative that the shooting was a hoax, which became a central tenet of his hugely popular InfoWars site. At the time of writing, HONR has successfully brought Jones to court on defamation charges, and Jones has become persona non grata as major social media sites deny him a platform. Even so, there remains an ample audience inherently distrustful of conventional sources, and it is nearly impossible to dislodge such beliefs, no matter how outrageous or easily debunked. In such circumstances, believers reassure one another that abuse of suffering people is not only acceptable but admirable.

The problem is that such dubious sources quickly become self-insulating echo chambers, utterly impervious to intrusions from reality that might otherwise keep them grounded. In this environment, the most odious of ideas can become a unifying gospel, and the most deplorable of actions rendered laudable. This is in no way unique to Sandy Hook truthers; all shades of conspiracy theory have similar bubbles, from the anti-vaccine movement to the believers in the moon-landing conspiracy. These cauldrons of conspiracy also exist in political discourse, with our self-selection bias skewing perceptions. And for all the social comfort that an affirming source might provide, such sources can be supremely damaging in the long run, especially when mental health is involved.

This is exemplified by the growing online presence of "Targeted Individuals" (TI) groups, a community of people convinced that an entire cabal of shady operatives are shadowing their every move. As evidence, TIs claim to hear voices in their head with sinister and disturbing messages. Their rationale for these discomforting experiences is that they are the victims of a covert government plot with "energy weapons," glorified lab rats in a mind-control experiment. The TI community has grown hugely in the previous few years, with adherents even setting up Mindjustice.org, a registered charity that seeks to prevent governments from using these hypothesized weapons. Across YouTube and the wider internet community, there are multiple accounts alleging an epidemic of "gang-stalking."

There is, however, a more mundane if no less sad explanation for these sensations: delusional issues and paranoid schizophrenia. The phenomenon is still poorly researched, but scientists Lorraine Sheridan and David James found that each ostensible case of gang-stalking revealed a severe delusional disorder in complainants. Advising these individuals to get help is a Sisyphean task—the TI community insists that victims should not visit mental-health professionals. Most tragically of all, they also make the cult-like claim that family members who suggest the problem is psychological in origin might be in on the deception and can't be trusted. The sheer abundance of sites reinforcing such notions is a considerable part of the problem. As Sheridan laments: "There are no counter sites that try and convince targeted individuals that they are delusional. They end up in a closed ideology echo chamber."

The afflicted congregate in dedicated forums and entirely self-insulating bubbles that cement delusion. This ultimately denies sufferers vital psychological intervention and can end in tragedy. On November 20, 2014, Myron May, a New Mexico prosecutor, walked into the library of Florida State University and indiscriminately opened fire, shooting three people before dying in an exchange with police. Immediately before his death, May had uploaded a suicide note to YouTube, describing the agony of being a "targeted individual," stating that the shooting was to raise awareness of the TI community. This came just over a year after Aaron Alexis shot 12 people dead and injured three others at Washington Navy Yard. Prior to the massacre, Alexis had complained of being a targeted individual, under assault by "extremely low-frequency electromagnetic waves." It's important to note that the vast majority of psychosis victims never become violent, but it is clear that mental-health issues are not helped by a choir of enablers dismissing the need for help.

The very crux of these issues is one of sources, and indeed of our own biases. While the internet should in principle allow us access to a range of ideas, our human tendency toward confirmation bias makes us vulnerable to selecting narratives that reinforce what we already believe rather than challenging the flaws in our own reasoning. And for those with their fingers on the pulse, there's no shortage of profit to be made from catering to people's prejudices and telling them precisely what they want to hear.

Take, for example, the rise in hyper-partisan websites. We might select two at opposite ends of the spectrum. *Liberal Society* features such clickbait headlines as WOW, SANDERS JUST BRUTALLY EVISCERATED TRUMP ON LIVE TV. TRUMP IS FUMING. This is clearly worded to attract left-leaning voters. At the other end of the spectrum, *Conservative 101* also exploits ludicrous premises to appeal to the right of the aisle, with hyperbolic headlines such as NANCY PELOSI JUST HAD MENTAL BREAKDOWN ON STAGE AND MADE CRAZIEST STATEMENT OF HER CAREER. On the face of it, these two websites couldn't be more different. So it may come as a surprise that they are both creations of the same Florida company. A 2017 Buzzfeed investigation found that some of their stories were worded almost identically, bar the headline and some choice words designed to stoke political fires in their respective audiences. The reason for this is entirely cynical: Hyper-partisan pieces reap much engagement on social media, and the more emotively the piece is framed, the more likely it is to be shared. And with rampant sharing comes increased advertising reach and profit for the parent company. In most cases these stories were taken verbatim from regular news outlets, modified for their respective audiences to increase their shareability.

This, sadly, is far from unusual. A 2016 *New York Times* investigation found that "political news and advocacy pages made specifically for Facebook, uniquely positioned and cleverly engineered to reach audiences exclusively in the context of the news feed" were the premier sources of political engagement, despite their shaky connection to reality. Not that this is unexpected, of course; as we've seen, people have a predilection for finding ideological echo chambers and there's ample money in telling people something that confirms their prejudices. The bombastic headlines and emotive hyperbole native to such sites is not accidental either. The product they pitch is outrage, which in turn breeds engagement. They are glorified outrage machines, designed to coax profit from anger.

While exploiting political polarization for profit is exceedingly cynical, what is infinitely more alarming is the ease with which vested interests can manipulate sources or conjure them out of thin air. During the 2016 US presidential election, the fingerprints of Russian intelligence were all over much of the partisan disinformation about Hillary Clinton. Quite aside from targeted leaks, an in-depth investigation revealed that anti-Clinton propaganda was tailor-made for receptive audiences across the political spectrum. Propaganda measures targeting the rabid alt-right were plentiful, but equally prominent were similar measures aimed at the left that painted a false equivalence between Clinton and Trump, prompting a section of the left to become useful idiots under the mantra of "Never Hillary!"

Odiously, these seemingly divergent narratives were orchestrated by Russian intelligence, crafted to undermine Clinton's campaign. A 2015 investigation by *The New York Times* found that a veritable army of Russian trolls were employed to impersonate pro-Trump Americans on social media and to spread anti-Democratic Party conspiracy theories. By 2016, the CIA had clear evidence of Russian tampering, and in January 2017, the joint US intelligence agencies reported that Russia had conducted a massive cyber operation on Putin's order. The stark conclusion was that the Russian administration had acted to influence the election in Trump's favor by discrediting Clinton. To do so, Russia had conducted a sustained campaign, targeting hacking of the Democratic Party and extensive propaganda across social media and Russian-controlled news platforms. In particular, the Russian TV channel RT in the US was singled out as being a "messaging tool" for the Kremlin.

At the time of writing, this scandal is still unfolding and the ramifications remain unclear. But Russia's mastery of such subterfuge shouldn't be surprising. *Dezinformatsiya* (disinformation) has a long pedigree, but its industrialization emerged with the modernization of media and mass communication. This is reflected in the etymology of the word itself, which arose independently in both Russian and English to characterize the spread of propaganda across Europe in the prelude to the Second World War. Russia quickly recognized the sheer potential of this modality and, as early as 1923, the GPU (forerunner to the KGB) had established an office dedicated to this purpose.

This quickly became an integral part of Soviet intelligence and, by the dawn of the KGB in the 1950s, it became an essential component in the doctrine of "active measures," the art of political warfare. The remit of active measures was wide-ranging, and included media manipulation, front groups, counterfeit documents, and even the occasional assassination when required. It became the beating heart of Soviet intelligence. KGB Major General Oleg Kalugin described it as "not intelligence collection, but subversion: active measures to weaken the West, to drive wedges in the Western community alliances of all sorts, particularly NATO, to sow discord among allies, to weaken the United States in the eyes of the people of Europe, Asia, Africa, Latin America, and thus to prepare ground in case the war really occurs."

Throughout the Cold War, the Soviets were virtuosos in creating tensions between allies. They exceled at black propaganda, crafting damaging material that purported to be from the other side, planted to ensure wide propagation. "Operation Neptune," for example, was a coordinated attempt in 1964 to use forged documents in order to imply that prominent Western politicians had supported the Nazis. This was quickly exposed as a counterfeit, but other ruses were more successful.

The chief target of *dezinformatsiya* was, unsurprisingly, the United States. Curiously, the US was far behind in the propaganda stakes. Russian efforts were largely ignored in the US until 1980, when a Soviet forgery of a presidential document claimed that the administration supported apartheid. This bogus claim got traction in US media, so appalling President Jimmy Carter that he demanded a CIA inquiry into Russian disinformation. The ensuing report found ample evidence of Russian-planted falsehoods throughout the world, including counterfeited documents suggesting that America would use nuclear weapons on her own allies. Not that this report stopped the flow; when Ronald Reagan sought a second term, KGB Chairman Yuri Andropov decreed that it was imperative that all KGB foreign intelligence officers, regardless of office, take part in active measures. In 1983, the KGB instructed its American operatives to move against Reagan's possible reelection, and stations the world over were ordered to popularize the slogan "Reagan Means War." Despite the best efforts of Soviet intelligence, he was reelected in a landslide.

But, although the Soviets failed in this endeavor, they had learned something powerful: Outright interference might be difficult, but undermining trust in their adversary was much more fertile territory. Conspiracy theories are a potent weapon; the KGB crafted elaborate conspiratorial narratives, circulating them in groups sympathetic to such views. The extent of this was revealed when KGB archivist Vasili Mitrokhin defected to the UK in 1992. The Mitrokhin archive is a treasure trove of information about how Soviet intelligence influenced public opinion by cleverly planting stories with vocal sources. Especially popular was a series of narratives that insisted John F. Kennedy had been assassinated by either the CIA or wealthy bankers. Naturally enough, this was planted—with exaggerated or falsified evidence—among conspiratorially minded groups and still persists in various forms to this day. The KGB saw the US's difficulties as their own opportunities and were quick to exploit America's racial strife, planting claims that Martin Luther King Jr. was an "Uncle Tom" agent of the US government and, in the wake of his assassination, perpetuating claims that he too had been assassinated by the CIA. Internal consistency didn't matter; what was crucial was sowing the seeds of discord.

The KGB also found a receptive audience for their claims that water fluoridation was a government plot for mind and population control, a notion that still has some currency in alternative circles. The charming irony of all this is that the conspiracy-obsessed groups exploited to disseminate fictions were completely unaware that they had become useful idiots in a very real plot.

Perhaps the most nefarious and potentially damaging of all was Operation INFEKTION, an attempt by Soviet intelligence to claim that AIDS was a man-made virus, manufactured by the US government to control the population of "undesirables." We've already alluded to this belief, but it's worth expanding on precisely how it spread. The first incidence was in 1983, when an Indian pro-Soviet newspaper carried an anonymous letter from a "well-known American scientist," claiming AIDS had been developed in a secret biological weapons laboratory in Fort Detrick in Frederick, Maryland. When the claim was largely ignored, Soviet intelligence realized the problem was that the source was simply not strong enough.

To remedy this, they relaunched the story in 1985 as the AIDS crisis was intensifying, this time enlisting a pseudoscientific report by retired biophysicist Jakob Segal. This report claimed that the AIDS virus had been synthesized by combining parts of other retroviruses, VISNA and HTLV-1. Segal further claimed that the military had tested the virus on prisoners, resulting in its spread. Although an East German, Segal masqueraded as French to hide his communist affiliations. The pitch had the desired effect; by 1987, it had received coverage in more than 80 countries in 30 languages, primarily in leftist publications. The dissemination followed a well-established pattern; the story would appear in a publication from outside the USSR, and later be presented in the Soviet media as the investigative work of others. To explain how AIDS was so prevalent in Africa, Radio Moscow claimed that a vaccination project in Zaire was in fact a deliberate attempt to infect Africans.

Initially at least, Operation INFEKTION was a success: Not only did it focus outcry against the US, it also distracted from Russian chemical and biological weapons development. But, while INFEKTION was an intelligence success, it was a pragmatic failure. The problem with narratives is that reality doesn't care much for them—AIDS was very real and Russia had no special immunity. The virus began wreaking havoc there in the mid-1980s, prompting Soviet virologists to seek help from their American colleagues. It was made clear through diplomatic channels that no help would be forthcoming until the disinformation campaign ceased. While the Gorbachev administration initially attempted to derail American attempts to expose the disinformation, they eventually grasped the scope of the problem, and in 1987 Operation INFEKTION was officially disowned. In 1992, Russian Prime Minister Yevgeny Primakov apologized, admitting the KGB had manufactured the story to sow seeds of discord in the US. Sadly, once a rumor takes root it continues to mutate and grow far beyond its initial constraints, and, as we have seen previously, AIDS denialism is very much alive.

Although Soviet *dezinformatsiya* greatly subsided following the dissolution of the USSR, the elevation of ex-KGB agent Vladimir Putin to high office meant that such measures have yet again become policy. Russian state media such as *Russia Today* are de facto propaganda channels, but more recently Russian intelligence has embraced "troll factories," where bloggers are paid to flood forums and websites with comments critical of the West and supportive of Putin, and to seed disinformation. The existence of such groups was thrown into sharp relief following Russia's annexation of Crimea in 2014, when anti-Kyiv sentiment began to appear on an industrial scale, as well as posts critical of Barack Obama and NATO that were ultimately traced back to troll factories in St. Petersburg.

The effectiveness of Putin's subversion will doubtlessly be long debated, but it is a tangible reminder of the dangers of dubious sources. It's worth noting that Russia's attempts to undermine opponents is not limited to the far side of the Atlantic. These divide-and-conquer propaganda pitches extend far beyond the US and include attempts to influence the 2017 French election, campaigns to stoke anti-immigrant sentiment in Germany and Sweden, and some evidence that the cybercorps of Russian intelligence may have used similar tactics to sway the 2016 EU Brexit referendum. This is simply an old technique with new tools, the chief difference being that social media makes it so much easier to propagate a dubious belief than it was during the Cold War.*

The ongoing coronavirus pandemic is an exemplar of precisely this. With every nation in the world affected to varying degrees, the resultant fear created a vacuum too easily filled by malicious

---

* While Russia is the clear world leader in the field, other countries have certainly dabbled. In 2000 *The New York Times* alleged that the CIA had used fictitious stories planted in newspapers to try to influence Iranian politics. Under the Reagan administration in 1986, the US used disinformation to undermine Libyan ruler Muammar Gaddafi, prompting Bernard Kalb, spokesman for the State Department, to tender his resignation in protest over such tactics.

fictions. Perhaps most odious is the claim that coronavirus is a genetically engineered bioweapon, designed for havoc and depopulation.
Leaving aside the fact that COVID-19 would be an unreliable and
subpar agent of war, the astute reader will see that this is the same
false narrative that underpinned Operation INFEKTION decades
before. Unsurprisingly, the fingerprints of Russian disinformation
have already been found lurking beneath the seemingly organic sharing of this myth across social media. China too has borrowed heavily
from Russia's mastery of subversion, using a network of social media
accounts linked with government to propagate a flurry of conspiracy
theories. The aim is the same as it ever was: to sow discord among
enemies and deflect attention from their own failings.

This is a potent reminder that there are often vested interests and
hidden agendas behind the material we happen across, be it a political
motive or clickbait for profit. No matter how satisfying or shocking
the material might be, we should strive to verify information before
we rush to share it. Thankfully, there are some resources for this;
long-standing websites like snopes.com keep close tabs on many of
the regular offenders, and a quick visit to the site can stave off the
damage of an ill-supported share. There is also a growing number of
fact-checking resources online, covering everything from the claims
of alternative medicine to political rumors.

To avoid becoming vectors for misinformation and being fooled
ourselves, the motto of the Royal Society, *nullius in verba* ("take no
one's word for it"), is a useful one. It's a reflection of the fact that
the onus is on us to determine the truth of all statements before
we accept them, rather than simply adopting something because it
resonates with our sensibilities or provokes our ire. For in truth, we
are more swayed by coverage than we might be aware.

18

BAD INFLUENCERS
Celebrities, Wellness Cults, and
Pseudo-Profound Bullshit

There have been few downfalls quite as spectacular as that of health and lifestyle author Belle Gibson. Until 2015, the young Australian author was feted as an inspiration, a young woman who overcame multiple cancers, rejecting conventional treatment and wholeheartedly embracing alternative medicine and natural remedies. Despite a bleak prognosis, Gibson persevered even when the cancer metastasized to her blood, brain, and uterus; she remained steadfastly determined to survive. She suffered a stroke, even briefly dying on the operating table, and still she ignored the pessimistic predictions of her medical team. Eventually, against every conceivable odd, Gibson made a full recovery.

Such an incredible about-face in the face of a deadly situation made her a beacon of hope, cementing her celebrity. Hundreds of thousands of her online followers rejoiced at her success, and her incredible recovery was widely covered the world over. She was lauded by publications such as *Elle* and *Cosmopolitan*, the latter referring to her as "the most inspiring woman you've met all year." Her subsequent "wellness" application and health cookbook, *The Whole Pantry*, was eagerly anticipated. Publishing companies scrambled to woo her for the rights to her forthcoming tome, and Apple even flew her to California for the launch of their iWatch, which shipped with her app, racking up sales in excess of $1 million.

But Gibson's inspiring story glazed over many cracks, the extent of which would soon become apparent.

The first warning sign was the misappropriation of charity funds. Gibson had styled herself as a philanthropist and often casually talked about the huge funds she was raising for charity. Fairfax Media began to investigate in late 2014, finding that, although Gibson had solicited in the name of five charities, none of these groups had received any money. A further two groups had also had their names used, but the amounts donated did not tally with Gibson's claims. Of the A$300,000 Gibson had boasted about donating, only A$7,000 appears to have gone to any charitable causes—and most of that after the investigation began.

Embezzlement was only the beginning. Questions quickly piled up and Gibson's halo rapidly tarnished. Some began to treat her miraculous recovery with a skepticism that had once been entirely lacking. Under questioning, she was unable or unwilling to name her doctors. More than that, her surgical scars appeared nonexistent and the details of her story unraveled under even the gentlest questioning. Initially, Gibson crafted a tepid response, suggesting that perhaps she had been misdiagnosed. But indications of fabrication mounted and, confronted with this growing body of evidence and even the potential of criminal charges, Gibson admitted in December 2014 that her entire story was a fraud. She had never had cancer and there was no miracle recovery.

Inevitably, a furious reaction to such betrayal ensued. As Gibson was to discover, those who would place you on a pedestal are often the first to tear you down. The very media outlets who formerly praised her reacted to her deception with unadulterated outrage. But their shocked indignation would have been markedly more convincing had they not elevated her to sanctity in the first instance.

Without such fawning, uncritical praise, Gibson would have remained just another crank peddling the wellness narrative to an echo chamber of the similarly convinced rather than a mainstream hero and minor celebrity. Gibson's story is one of greed, arrogance, and hubris—but it was one aided and abetted by an uncritical media. Her entire narrative had been amplified by journalists who projected a heroic story arc upon her, failing to ask even cursory questions. Despite the sirens of bad science sounding from the outset, none of the publications lauding Belle Gibson bothered to check her claims—*Cosmopolitan* writer Lauren Sams defended their hagiography, stating: "Cancer is so all-consuming, so catastrophic, so final, that to question anyone's diagnosis would just be downright evil." But this is a poor excuse for shirking responsibility. It was hardly cruel to ask for more information on Gibson's condition or to request to interview her doctor.

The reason these questions were not asked is that Gibson fitted a marketable archetype, an inspirational figure with asinine platitudes that could be widely shared on social media. In short, Gibson was a celebrity-in-waiting. And in this, there's a subtle but important point about the impact of celebrity and popular perception. Her story fitted an easily sold narrative. But the lack of due diligence on the part of the outlets that elevated her to the mainstream cannot be overlooked. The warning signs were there long before the deception was revealed. Gibson claimed to have contracted cancer due to a reaction to the cervical-cancer vaccine Gardasil, an entirely bogus claim that ought to have been a huge red flag. Moreover, Gibson explicitly rejected conventional cancer therapy, condemning it as unnatural. Instead, she attributed her cure to a fruit diet, Ayurvedic medicine, craniosacral therapy, colonic irrigations, and Gerson treatments—all of which are pseudoscientific with no demonstrable efficacy against cancer.

The unquestioningly positive coverage ignored all this and was tantamount to endorsement of these dangerous, untrue assertions. While many people were understandably aghast at Gibson's cancer deception, what was far more concerning was the dreadful medical advice she gave and the staggeringly uncritical media coverage this dangerous advice was afforded. What is downright unethical is the claim that conventional cancer treatment can be neglected in favor of alternative remedies and natural eating, which is precisely what Gibson preached and numerous publications propagated by extension. This would have been deplorable even if Gibson had had cancer, given that there would still not be an iota of evidence that her remission was caused by any of the agents credited.

Belle Gibson is in no way unique; there are thousands like her who extol the virtues of pseudoscientific alternative medicine or diet and are happy to profit from it. If celebrity is something of a nebulous term, "influencer" is an even more amorphous, vaguely dystopian concept; one fittingly ominous definition is a "third party who significantly shapes the customer's purchasing decision but may never be accountable for it," an aspirational figure with precisely zero responsibility. The burgeoning "wellness" industry is rife with influencers selling this aspirational version of health, encompassing everything from weight-loss treatments to spas to alternative medicine. In totality, it's valued at $4.2 trillion—almost four times the global pharmaceutical industry. Unlike pharmaceuticals, many of these elements have no evidence of efficacy; alternative medicine alone accounts for over $360 billion of the total figure. Operating under the vaguely new-age concept of "wellness" as a holistic mind/body/spiritual form of health, advocates subscribe to a simplistic "natural" philosophy, embracing widely debunked alternative medical treatments and theories. Assertions made by advocates might be fictitious, but they garner ample publicity. Some are well-meaning if misguided believers, genuinely convinced that the remedies they hawk are effective, even when faced with ample evidence to the contrary. Others still are morally bankrupt charlatans who easily shift their snake oil onto a gullible market.

This cult of wellness is well represented on the internet, with gurus sharing nonsense to enrapt audiences. Some of these gurus are worryingly devoid of critical-thinking skills. Vani Hari, operating under the moniker "the Food Babe," is one such figure who profits from making demonstrably false claims about food safety. Sometimes this yields unintentional hilarity, such as her spectacular insistence, in 2014, that airplane air is dangerous as it's not pure oxygen and has a nitrogen content of "almost 50 percent." Readers with a high-school education might recall that our atmosphere is in fact 78 percent nitrogen and that pure oxygen would quickly damage us irreparably. This howling error epitomizes Hari's ignorance of topics she pontificates on, compounded by a chronic inability to use Google.

Hari is far from alone on the digital wellness front. Naturalnews.com is a microcosm of this market, a portal for all manner of alternative claims, from hawking dubious dietary supplements and alternative medicine to propagating grandiose conspiracy theories. Its founder, Mike Adams, has adopted the nom de guerre "Health Ranger," with nary a hint of irony or self-awareness. Like many sources in the alternative health field, Naturalnews has an extremely low view of best medical evidence, and a poor ratio of fact to fiction. The demonization of benign and beneficial public-health measures, tinged with accusations of conspiracy, is a recurrent theme in such spheres, and Naturalnews is no exception. Fearmongering is rife, with a bingo card of long-debunked claims, from "vaccinations cause autism" to "fluoride was used by Nazis for mind control." Such stories perpetuate with furious tenacity, impervious to any intrusion by reality.

Unsurprisingly, conventional medicine is viewed with suspicion, and hysterical ill-founded claims of "Big Pharma" conspiracy arise with regularity. Like other sites of its class, Naturalnews is also home to all manner of bogus cancer claims and outright misinformation. In the wake of Patrick Swayze's death from pancreatic cancer, Adams declared that Swayze "joins many other celebrities who have been recently killed by pharmaceuticals or chemotherapy," a deplorable comment betraying a staggering ignorance of cancer treatment.* For Adams, conspiracy lurks everywhere; in addition to his stance on health issues, he is an AIDS denialist, a 9/11 truther, and an Obama birther. Unsurprisingly perhaps, he is also deeply opposed to genetically modified food and is currently under investigation for threatening violence against GMO researchers.

While it's tempting to dismiss Adams as a fringe figure with a penchant for seeing dark forces at play everywhere, to do this would

* Swayze himself had no tolerance for snake oil. In an interview before his death, he expressed his irritation at the nonsense he'd been hawked: "If anybody had that cure out there, like so many people swear to me they do, you'd be two things: You'd be very rich, and you'd be very famous. Otherwise, shut up." We should all be crazy for Swayze.

underestimate the powerful allure of the brand; Naturalnews pulled in an impressive 7 million unique views a month in 2015, with stories originating on the site traveling far and wide. This coverage is not by any means negligible, but it is not the biggest natural-health site on the web. That dubious distinction belongs to Joseph Mercola of mercola.com, and his far-reaching empire of nonsense.

In many respects, Mercola relies on the same pseudo-naturalistic spin as other players in this field; he is a vocal opponent of vaccination, with a track record of dangerous fearmongering on the topic. He too has suggested that HIV does not cause AIDS and has lionized Tullio Simoncini, a man who claims that cancer is a fungus and baking soda its cure. On top of this, Mercola promotes widely debunked therapies such as magnetic healing, homeopathy, and psychic medicine. Like others of his ilk, he sells a range of alleged wellness products, at least four of which have led to warning letters from the FDA for illegal claims.

Despite the condemnation of "Big Pharma" greed by Mercola and his contemporaries, they are not so keen to acknowledge the fact that selling dubious cures is remarkably profitable. Mercola runs a slick outfit that shamelessly employs pseudoscience to sell useless or potentially dangerous products, and for this he is well compensated. A damning review in 2006 by *BusinessWeek* was exceptionally critical of his aggressive marketing and "lack of respect" for visitors, noting that "he is selling health care products and services, and is calling upon an unfortunate tradition made famous by the old-time snake oil salesmen of the 1800s." This does not seem to have deterred his customers one jot, and Mercola took in a tidy $7 million in 2010 alone.

Were these purveyors of nonsense confined solely to internet fringe groups, it would be bad enough, but all too frequently they are given an air of respectability by mainstream sources. The Belle Gibson debacle and the fawning coverage she received before her fall from grace exemplify the reality that the cult of wellness is a mainstream phenomenon. Gibson is far from alone in this respect;

the Food Babe's debut publication, *The Food Babe Way,* hit the number one spot in both *The New York Times* and *Wall Street Journal* bestseller lists when released in 2015.

You may point out that I'm extrapolating from a certain amount of presupposition. So what if the media coverage that such a minor celebrity garnered was inaccurate; surely this can't reasonably be expected to sway anyone? Does it really matter to public understanding if pseudoscientific outfits hawk nonsense to their audience? These are understandable objections. It sounds outlandish to suggest that the vapid musings of some social media personality would have any impact on public health, and certainly not the quasi-philosophical ramblings of someone more concerned with virality than profundity. Yet evidence to the contrary exists, stemming from a seemingly unlikely medium: reality television.

Even among devotees, reality television does not enjoy a reputation as especially profound programming. It is, by definition, voyeuristic, concerning itself primarily with the petty. To stoke interest, producers aren't above hiring abrasive or flamboyant personalities to titillate viewers, or manipulating events so that the worst possible traits are amplified. It's notable too that there is an air of derision about many of the contestants. One might be forgiven for thinking they're selected as objects of mockery, unwitting fools to whom the viewing public can feel superior. Jade Goody, a contestant on the 2003 UK version of *Big Brother,* exemplified this role perfectly.

Throughout the tabloid press and viewing public, Goody's lack of basic general knowledge became a source of constant mirth. Among her many faux pas, she thought Rio de Janeiro was a person, and was unaware that English was spoken in America. Despite the plentiful mocking headlines, Goody's brash personality and occasionally ridiculous behavior saw her become a regular fixture in tabloids and vacuous gossip columns. But in late 2008, the tone shifted when she was diagnosed with cervical cancer. After initially optimistic assessments, Goody and her family were told the cancer was in fact advanced. By March 2009, Goody was dead, leaving behind two small children. She was 27 years old.

What followed was remarkable. Until that point, the number of younger women getting Pap smears had fallen to a worrying low in the UK. Following the coverage of Goody's death, however, this trend rapidly reversed. In March 2009, appointments surged by more than 70 percent above projected figures, and it was estimated an extra 500,000 screenings took place between Goody's diagnosis and death. While "reality TV star" might be considered one of the weakest formulations of the word "celebrity," the effect was tangible. It was also unduly pronounced in women from lower socioeconomic backgrounds, who are often missed by health campaigns.

Such screening is likely to have saved lives and there is no doubt coverage of Goody's illness had a direct impact. But, just as celebrity is often fleeting and transient, a similar caveat applied to the positive impact of this coverage: a story only has emotive power for as long as it's in the public consciousness. Once interest wanes, memories fade, with the half-life of our interest directly proportional to media coverage. The Goody effect was no exception—the impressive rally of screening numbers in 2009 coincided with peak coverage, but by 2012 the numbers were rapidly falling away. By 2017—once memory of Goody's plight and the surrounding coverage had waned—cervical cancer screenings in the UK had fallen to a 19-year low.

What is also important to note is that it is celebrity coverage itself that impacts screenings rather than any intrinsic value or benefit of these screenings. Requests for mammograms more than doubled following the wide coverage of Kylie Minogue's breast cancer scare in 2005. But most of these requests came from young women, for whom screening was likely to produce false positives and hence be detrimental. Nor did the increased coverage equate with improved understanding; as with many cancers, age is the strongest risk for breast cancer, and the vast majority occur in older patients. As

Lesley Walker of Cancer Research UK cautioned, skewed coverage "may also set up a chain of panic among young women, while misleading older women to think that ageing is not a relevant factor in breast cancer."

Curiously, there is a historical media precedent for the misconceptions that abound around breast cancer. In 1999, the German magazine *Stern* ran a feature claiming that one in ten women would get breast cancer in their lifetime. The article came with emotive vignettes from young breast-cancer survivors, with the topless survivors baring their breasts and mastectomy scars. The powerful feature was a hit, syndicated far and wide in a host of different languages. Rapidly, the one-in-ten figure became branded into collective consciousness, used at the forefront of breast-cancer awareness campaigns.

Yet this figure, while technically true, was completely misleading. *Stern* neglected to mention that the figure quoted referred to one's cumulative risk of having breast cancer by age 85—by which time, victims are far more likely to have died from other illnesses. While average age at diagnosis was around 65, the mantra of the article caused young low-risk women to massively overestimate their risk, with one study suggesting a twentyfold overestimation. There's ample evidence too that this skewed perception is a result of coverage bias—while physicians surveyed understood 65-year-olds were more at risk than 40-year-olds, only about 20 percent of women surveyed were aware of this fact.*

But for the raw power of celebrity on perception, there's no figure in modern history quite as influential as Oprah Winfrey. Her eponymous show ran for 25 seasons, rendering her the voice of America. Her influence on public opinion, dubbed the "Oprah effect," was unprecedented and unparalleled. In everything from music to publishing, Oprah's imprimatur was without equal; books recommended on her show became overnight bestsellers and

---

* The specific misconceptions about breast cancer and illness are covered in detail in Gerd Gigerenzer's excellent book *Reckoning with Risk*.

guests became celebrities. By one estimate, Oprah's endorsement of Barack Obama in the 2008 presidential election was thought to have yielded him a million extra votes in the primary. Even today, through the Oprah Winfrey Network, her ability to touch and inspire remains. Winfrey is intelligent, liberal, and very outspoken—and to her immense credit she has raised issues that other shows steered clear of, bringing important topics to public attention.

Despite these admirable qualities, however, Oprah's penchant for pseudoscience has given platforms and a veneer of respectability to some truly odious ideas. Throughout the show's long tenure, the medical advice frequently given on Oprah was demonstrable tripe. And worse, that coverage itself gave an air of respectability to people with vested interests and dangerous wares that should not be entertained. Surgeon and writer David Gorski, while praising her talent, lamented that "unfortunately, in marked contrast, Oprah has about as close to no critical thinking skills when it comes to science and medicine as I've ever seen. . . . No one, and I mean no one, brings pseudoscience, quackery, and antivaccine madness to more people than Oprah Winfrey does every week."

Many of Oprah's guests push long-discredited theories on health issues. Anti-vaccine campaigner and *Playboy* model Jenny McCarthy was a regular guest and given a regular column in Oprah's bestselling magazine *O* and on her website. Even the medical guests propagated medically unsupported views. Christiane Northrup, for example, told an audience member on *Oprah* that the HPV vaccine could kill and advised against it. It is highly likely that many of Oprah's viewers might have been unaware that this claim stands in stark opposition to recommended medical advice, or that Northrup claims astrology and tarot cards are diagnostic tools. Such irresponsible advice is potentially fatal, yet it went completely unchallenged.

The show was a melting pot of feel-good new-age philosophies and often meaningless platitudes. For example, *The Secret* was praised on Oprah's show as a revolutionary tool. The central tenet of the book was the "law of attraction," which claimed that "focused positive thinking can have life-changing results such as increased wealth, health, and happiness." Oprah's recommendation helped catapult it onto *The New York Times* bestseller list, where it remained for 146 consecutive weeks. The show extolled the power of positive thought. Of course, the converse of such ideas is the implication that people in bad situations simply haven't thought positively enough. Taken to the extreme, such logic implies that starving children and those trapped in war zones have only themselves to blame.

Such framing came with consequences. In 2007, Kim Tinkham forwent conventional treatment for her breast cancer and instead opted to rely on positivity and alternative medicine. While her doctors pleaded with her to get conventional treatment that would almost certainly have saved her life, she instead appeared on Oprah's show, lauding her new treatment with Robert O. Young. In Oprah's defense, she expressed concern in the segment that this was perhaps taking positive thinking too far. This was an under-statement—Tinkham's treatment pivoted on the discredited notion that cancer was caused by excess acidity. Young eventually claimed she had been "cured" of cancer, and included a testimonial from her on his website. Kim Tinkham died of the same breast cancer in December 2010, at the age of 51. Robert O. Young eventually ended up in prison for practicing medicine without a license.

Oprah's defenders may argue that these are not "her" views but rather those of her invited guests. Yet Oprah herself frequently endorsed the viewpoint of her guests, shielding them from criticism on the show. She also championed several of these ques-tionable beliefs; when actress and self-help guru Suzanne Somers was on air hawking "bioidentical hormones," Oprah lavished praise on her and used the medically unsound procedure herself. She even criticized the medical profession for not giving this profoundly dubious approach the respect she felt it deserved, opining that "we have the right to demand a better quality of life for ourselves . . . and that's what doctors have got to learn to start respecting."

Oprah's protégé Mehmet Oz is arguably worse. *The Dr. Oz Show* commands a typical viewership of 4 million worldwide in addition to a range of magazines and books, rendering him arguably the most famous medical professional on the planet. While some of the advice given on his show regarding diet and exercise is reasonably sensible, it is buttressed by outrageous remedies and claims that fly in the face of scientific evidence. Oz is a keen proponent of alternative medicine, and gives ringing endorsement to homeopathy, psychics, and highly dubious diet supplements.

His show provides a platform for purveyors of all manner of dubious medical advice and products such as Mercola, who is lauded by Oz as a "pioneer in holistic treatments." He has also given free rein to master of quantum quackery Deepak Chopra and lavished praise on the Food Babe. Questionable guests and dubious promotions aside, much of Oz's medical advice is frankly terrible; a study in the *British Medical Journal* found that 51 percent of medical claims on the show were either not supported by scientific literature (36 percent) or actively contradicted by best evidence (15 percent). While Oz commands a devoted audience, it has come at the cost of his professional reputation. He has been awarded James Randi's Pigasus Award on multiple occasions, a dubious honor reserved for those promoting outright quackery. In 2015, physicians from across the United States sent a letter to Columbia University condemning the fact that he had a faculty position, accusing him of "an egregious lack of integrity by promoting quack treatments and cures in the interest of personal financial gain."

Despite the ire that Oz provokes in his fellow professionals, the impact of their protestations is unfortunately limited. Unlike Oz, most doctors do not have a media empire at their disposal. No matter how well considered his critics' rebukes might be, they cannot hope to amplify their signal enough to counteract such noise. Sadly, it is often the case that coverage alone is enough to cement discredited ideas, rendering them practically immune to criticism. The buoyant market for "cleanses" that promise to "detox" consumers is a prominent example; these are completely useless from a scientific standpoint. We already have much cheaper and more effective tools for this in the form of a functioning liver and kidneys, which filter toxins quite admirably. Detox products simply don't work. As Edzard Ernst laments, the term has been "hijacked by entrepreneurs, quacks, and charlatans to sell a bogus treatment."

Still, detox products and cleanse diets top $5 billion in sales annually, driven to a large extent by celebrity endorsement. Dietary regimens of female celebrities in particular are touted to excess in both print and modern media. Pop star Katy Perry, for example, credited her glowing appearance in *Vogue* magazine to a cleanse diet.* The rich and famous have outsized influence on public perception, due in part to aspirational values we project onto those we deem successful, and hope to emulate—even when the wares they endorse are totally useless. The actor Gwyneth Paltrow takes things to an extreme with her lifestyle newsletter *Goop*, offering cleanses, exotic lifestyle products, and curated supplements at eye-watering prices. Typically these veer into complete pseudoscience; so frequent are Paltrow's bogus claims that gynecologist and science writer Jen Gunter has a section on her site expressly dedicated to pointing out the utter inanity and potential harm of nonsense claims made on *Goop*, from vaginal steaming to claims that bras cause cancer.

The jade egg is but one example. Priced at $66, *Goop* recommends women carry this golf-ball-sized stone in their "yoni" to "increase chi" and "stay in shape." To this, Gunter states: "Let me give you some free advice, don't use vaginal jade eggs." Despite the criticism from experts, *Goop* simply doubled down in an asinine post entitled "12 (More) Reasons to Start a Jade Egg Practice," presumably because the alternative title of "A dozen (stupid) reasons to stick rocks up your vagina" wasn't quite the marketing gold they

---

* While detox diets are useless, they shouldn't be confused with detoxification in the form of chelation therapy, used to treat patients exposed to dangerous amounts of toxic metal. In these treatments, a chelating agent binds to the metal in question to reduce harm. This shouldn't be confused with the colloquial form, but presumably Katy Perry is an unlikely candidate for heavy-metal poisoning.

were aiming for. Despite the absurdity of the notion and the chorus of experts expressing their concerns, none of this had any appreciable effect on sales, and jade eggs sold out rapidly. With incredible reach, celebrities and influencers can cause substantial damage to public understanding, especially when they stray into areas they do not learn about first. As the coronavirus pandemic rages, celebrities have too frequently amplified pseudoscience to massive audiences.

While there is no shortage of celebrities who confound public understanding of science and medicine, it needn't be this way. Historically, celebrity influence has sometimes been a force for good. In 1956 Elvis Presley was vaccinated against polio at a press conference, encouraging the under-vaccinated teenage cohort to get the intervention. Roald Dahl, who lost his daughter Olivia to measles, penned a heart-breaking and influential account in support of the vaccine. Actors such as Alan Alda have done fantastic work in advancing the public understanding of science. For better or worse, celebrity bestows a platform and the power to influence people. Whether this is employed for good or ill depends on the celebrity in question.

As an aside, the utterances of alternative gurus tell us something interesting too. David "Avocado" Wolfe is a case in point, pushing inspirational memes, anti-vaccine conspiracies, and suspect supplements to his 12 million Facebook followers. Wolfe distinguishes himself from a plague of similar hucksters by the sheer absurdity of his proclamations. For example, he insists that "chocolate is an octave of sun energy," a meaningless word salad typical of the milieu. Such statements are described by a delightful academic term: bullshit. Philosopher Harry Frankfurt defined bullshit as something designed to impress without any direct concern to the truth. *Pseudo-profound bullshit* is endemic in the wellness community. Researchers found that simply concocting impressive-looking statements with valid syntax and fancy terminology is enough to fool many, even when devoid of any intentional meaning. "Wholeness quiets infinite phenomena" was deemed especially profound by many surveyed.

The disconcerting reality is that mere exposure to claims makes us more likely to accept them. Ideas aren't so much exposed as absorbed. Falsehoods quickly take root. It is simply not enough for modern or traditional media outlets to hide behind the pretense of good faith while running destructive stories, nor can they shirk their responsibilities after the fact. And of course, as we increasingly curate and share our own media, we also share culpability.

As we've seen, the reality is that anecdotes and testimony hold disproportionate sway over us. In the cacophony of modern media, it can be extremely difficult to parse fact from fiction. This sounds grim, but a modicum of critical thinking can prevent such exploitation and anguish. When confronted with a claim, we should first question how reliable the source might be, and whether the source stands to personally gain from those claims. As a rule of thumb, we ought to be extremely wary of simplistic, reductive claims on complex topics such as health, politics, and science. Dramatic claims and alluring promises should be viewed with extreme skepticism unless proof is offered. In general, if something sounds too good to be true, it likely is.

It is crucial to demand that those making claims provide reliable evidence for their assertions. This unfortunately places an onus on consumers to be wary, but we are not alone. Reputable professional bodies offer advice on identifying dodgy claims—for example, the charity Sense About Science run a laudable "Ask for Evidence" campaign, supporting people to query claims on everything from health care to public policy. Simply learning to suspend the acceptance of a particular narrative until it has been independently confirmed is a hugely beneficial habit we can all adopt. There is good evidence that analytical thinking reduces acceptance of pseudo-profound bullshit, and encouraging people to reflect rather than to intuitively accept a statement makes them far more likely to spot dubious sentiment. If we desire to cut through the noise we're assaulted with and make sense of the world, we need to attune ourselves with analytical thinking and the tools of scientific skepticism. To do that, we need to explore what science is—and what it is not.

PART 6

The Candle in the Dark

What Science Is and What It Is Not

"Science is a way of trying not to fool yourself.
The first principle is that you must not fool
yourself, and you are the easiest person to fool."

—RICHARD FEYNMAN

THE EDGE OF SCIENCE

*Occam's Razor, Homeopathic Delusions, What Science
Really Is, and How to Spot Pseudoscience*

N*ature* is the most prestigious scientific publication in the
world. That which makes the hallowed pages of this long-
running journal captures the attention of the scientific community.
In 1988, an astonishing claim by a French immunologist resonated
far beyond academia. Jacques Benveniste claimed to have diluted
human antibodies to such extremes they were entirely absent. And
yet, an immune reaction was still seen, provided the solution was
vigorously shaken. To Benveniste, this was proof that the structure
of water somehow remembered what it previously contained. In
his words, it was akin to "agitating a car key in the river, going
miles downstream, extracting a few drops of water, and then
starting one's car with the water." Some called the phenomenon
"the memory of water," but it had a much older name: homeopathy.

Proposed by Samuel Hahnemann in 1807, the central dogma of
homeopathy is that the more dilute the remedy, the greater its
potency the antithesis of the scientific observation that the potency
of a solution is proportional to the concentration of active ingredient.
Homeopathic dilutions are incredibly extreme, typically at a dilution
of 30C—akin to having one active particle per 1 million billion billion
billion billion billion billion particles. Such dilutions wouldn't even be
possible on Earth,* so homeopathic solutions cannot contain active
ingredients. Proponents argue that this is irrelevant because water
"remembers" that which was diluted in it. But water has a "memory"
of only about 50 femtoseconds—a few billionths of a second.†

---

* For a previous paper, I calculated that to contain even one molecule of an active ingredient, a 30C
dilution requires a "planet" of water 15,000 times the mass of the sun and 28 times its size.

† Given that all water on Earth is part of a closed system and every molecule in existence has
undoubtedly seen quite an amount of effluent, we should perhaps be grateful for this liquid amnesia.

Physical impossibility aside, clinical data does not suggest any real effect exists. Hahnemann himself could be forgiven for his outlandish idea, as the existence of atoms was not proven until almost a century later. But given our knowledge of modern chemistry and physics, it seems perverse to hold a position so clearly divorced from reality. This ought to have sounded a death knell for homeopathy decades ago, yet Benveniste's new result offered a stark ultimatum: Either it was wrong, or everything we knew about physical science would have to be rewritten.

This left *Nature* editor Sir John Maddox in a bind. A physicist by training, he knew homeopathy had no plausible mechanism of action. But scientific integrity demanded that evidence not be disregarded simply because it jarred with conventional thinking. Benveniste's work had been subjected to peer review and, though skeptical of the result, reviewers found no obvious flaws in methodology nor any telltale signs of bad science. If Benveniste was right, he had discovered something revolutionary that demanded attention. Maddox's compromise was to publish with a "Note of Editorial Reservation," subject to the condition that independent investigators would oversee replication of the experiment. Even with this caveat, the article generated sensational headlines worldwide. To proponents of alternative medicine, this ostensible vindication by science's most illustrious journal was a stinging rebuke to the scientists who had long decried their beliefs.

Basking in the glow of press attention, the charismatic Benveniste became a celebrity. Maddox meanwhile assembled a team of investigators to aid him in the replication process. Chemist Walter Stewart had a track record of exposing scientific fraud and was a natural choice. The final member of the team needed to be someone uniquely adept at spotting trickery and deception. Maddox opted to choose not a scientist, but a magician: James Randi.

"The Amazing Randi" was a consummate entertainer who had practiced magic for decades, even touring with Alice Cooper in the 1970s. He was also an accomplished escapologist, besting Houdini's record for escaping from a submerged coffin. And, like Houdini, Randi had a passion for debunking charlatans and a well-deserved reputation for exposing trickery. His inclusion made sense; editorial reservations in *Nature* were extremely rare events. Only one had ever been published before, for a 1974 paper that claimed evidence for psychic ability in the Israeli performer Uri Geller. It was Randi who had demonstrated that Geller's feats didn't require supernatural powers, only sleight of hand and gullible investigators.*

Enthused, the press dubbed the trio "The Ghostbusters" as they set off for Paris. There, Benveniste insisted that scientist Elisabeth Davenas demonstrate because of her knack for getting the experiment to work. A number of vials were subjected to the experimental protocol, some containing plain water "controls" and others homeopathically treated versions. As before, homeopathic solutions triggered inexplicable activity. But the demonstration highlighted some concerning aspects. Most concerningly, it was unblinded—Davenas always knew whether her sample was a control or not. And that meant that, either subconsciously or deliberately, bias could creep in. To circumvent this, Maddox had his team "blind" the experiment. Labels were removed from samples, and Stewart engineered a secret code to distinguish controls from active samples. This code was placed in an envelope and, after Randi sealed the room from intruders, wrapped in tinfoil and hidden in the ceiling for good measure.

Precautions in place, the experiment was rerun on the unlabeled samples. To lighten the tense atmosphere, Randi entertained those present with sleight-of-hand tricks; this did not endear him to Benveniste, who resented the magician's presence especially. With results due to be announced at a dinner with the press, Benveniste had magnums of champagne resting on ice for his vindication. As the experiment concluded, the code was retrieved to unscramble the findings. A hush of excitement fell across the gathered

* Incidentally, Randi awards the annual Pigasus mentioned earlier to paranormal frauds. The first version of this was called the "Uri trophy."

researchers and press as the analysis arrived. To the chagrin of the French researchers, it could not have been more disappointing. Under the blind conditions, the astounding result simply did not manifest. The original paper was anchored to delusion, a pronouncement that reduced many present to tears.

The subsequent report by Maddox's team revealed further multitudinous shortcomings, including lab books betraying abysmal statistical methods and repeated cherry-picking. The damning report refrained from alleging fraud but questioned the potentially malign influence of undeclared funding from homeopathy giant Boiron. The sad reality was that Benveniste's group were such ardent believers in homeopathy that they had allowed themselves to be misled, having "fostered and then cherished a delusion about the interpretation of its data."

This is an exemplar of *pathological science*, where researchers are seduced toward false results by the siren song of wishful thinking—a scientist's version of motivated reasoning. Tellingly perhaps, Benveniste opted to indulge his rhetorical flair rather than admit honest error, slamming Maddox's investigation as "Salem witch-hunts and McCarthy-like prosecutions." In a dramatic flourish, he compared himself to Galileo, oblivious to the crucial difference that Galileo was vindicated by experiment, whereas numerous labs were unable to reproduce Benveniste's assertions. There are still a number of proponents of this utterly pseudoscientific "science-based homeopathy"—a stellar example of a phrase that is both oxymoronic and regularly moronic.

The simple application of Occam's razor, however, might have saved Benveniste the embarrassment. This rule of thumb suggests that when faced with multiple explanations for an observation, the one that requires the fewest additional assumptions is most likely correct. To explain the result, one can either accept that: (a) all known physics and chemistry is largely wrong; or (b) the experiment was likely flawed. While (a) is not impossible, accepting it means having to explain why copious amounts of established data and theory are wrong. By contrast, (b) only requires a single experiment to be flawed. Occam's razor is simply a heuristic, like those encountered earlier, and is therefore not foolproof. Nevertheless, when confronted with multiple hypotheses, it provides a signpost toward the most likely starting point. The same principle applies to medical diagnosis, as commonplace explanations are generally more likely explanations for symptoms than are exotic conditions. Theodore Woodward's famous dictum to his interns was: "When you hear hoof-beats, think of horses, not zebras." Fittingly, "zebras" has become medical slang for unusual conditions.

The water-memory debacle highlights something important about the nature of science. Science and human inquiry have slowly replaced ignorance and fear of the world around us with knowledge and beauty. Modern medicine has enabled us to live longer, healthier lives, and science has endowed us with a deeper insight into the nature of the universe. But while the fruits of science underpin our world, there's a worrying disconnect between our reliance on science and our understanding of what it truly is. To many, the scientific method is a nebulous concept onto which they can project the prejudices they desire; religious apologists often insist that science is every bit as grounded in faith as their pronouncements. Subcultures like the anti-vaccine movement fail to distinguish the crucial differences between anecdote and evidence. Media outlets are so single-mindedly focused on presenting "both sides" that they often entirely fail to discern between emotive assertion and reliable evidence. Politicians and lawmakers are consistently flummoxed by the subtleties of causation and correlation, much to our collective detriment.

The great Carl Sagan worried that "we've arranged a global civilization in which most crucial elements profoundly depend on science and technology. We have also arranged things so that almost no one understands science and technology. This is a prescription for disaster." Sagan's lament is not hyperbole, nor is it inevitable. Improving public understanding of science and critical thought would be of huge benefit both to society and to us as individuals. But misconceptions abound; to many people, science is a mere collection of facts and figures, a compendium of banal trivia forced upon them in their schooldays by the lab-coated high priests of an arcane religion. However, as Benveniste's story demonstrates, scientists aren't infallible. They can be fooled by subtle mistakes, seduced by spurious results, or even be corrupt. We've seen too that not all studies are created equal: Some are well designed, careful to exclude confounding influences, while others are underpowered or conducted with inappropriate methodology.

It might seem impossible to know which results to trust, but the beautiful thing about science is that we need only trust the method. A study in isolation is simply a single data point. Ideally, it is accurate, but for various reasons it might be flawed. What really matters is the complete picture, the trends that emerge when results and analyzes are pooled. This is why, for example, evidence for human-mediated climate change or the safety of vaccines is so overwhelming: Data from thousands of studies and theoretical models all point to the same conclusion. Conversely, climate-change deniers or anti-vaccine activists who clutch at single or weak studies are being disingenuous; cherry-picked studies in isolation simply do not trump overwhelming evidence.

Science is not a collection of immutable facts or sacred dogma; it is a systematic method of inquiry. Scientists are not priests of this arcane knowledge who make pronouncements. One's perceived authority or accolades are ultimately irrelevant; the theories of the most celebrated Nobel laureate can be instantly overturned by the experiments of the humblest student. Reality cares not one iota for our prejudices and egos. Scientific knowledge is always provisional, and our acceptance of findings should be proportional to the strength of evidence offered. New discoveries constantly refine our understanding, and theoretical insights act as a compass for discovery, rendering science ultimately self-correcting.

Even titans must bow to evidence. The late nineteenth century saw a cascade of rapid advances in science. Ancient mysteries about our world seemed to be yielding their secrets at a furious rate, and Lord Kelvin (born William Thomson) was at the helm of many discoveries. A scientific colossus, Kelvin made huge strides in mathematical physics, thermodynamics, and electricity. He was knighted for his work on transatlantic telegraph communications, famous far beyond the scientific sphere. The Système International unit of temperature is even named in his honor.

Toward the end of the 1800s, a new problem came to the fore of scientific discourse: the question of the age of Earth. The foremost geologist of the era, Charles Lyell, argued that Earth was shaped chiefly by gradual processes rather than by the sudden calamities so popular in spiritual texts. On the combined strength of all available evidence, geologists proposed the hypothesis that features of Earth from volcanoes to earthquakes could be explained by simple geophysical processes occurring at a roughly constant rate. For this to be true, it required an extremely ancient Earth—hundreds of millions or even billions of years old. Nor was it only geologists who edged toward the idea of an ancient Earth to explain their data. In the first edition of *On the Origin of Species*, Charles Darwin had estimated that it would have taken 300 million years to erode the Weald, an expansive chalk deposit that covered swaths of southern England. Intrigued, Kelvin turned his singular intellect and the might of mathematics to determining the age of Earth.

Kelvin began by assuming that Earth started as a molten sphere of liquid rock. The surface of the planet, bounded to space, would reach constant temperature in negligible time. Below this surface, heat would slowly dissipate outward, a finding shown by the French physicist and mathematician Joseph Fourier in 1822. The prodigious Kelvin was extraordinarily well placed to apply this theory to the problem; at a mere 16 years of age he had clarified a number of Fourier's calculations. The heat equation explicitly defines how the system changes with time, allowing one to ascertain the time that has elapsed since the system began, assuming the molten-sphere model. Being meticulous, Kelvin obtained estimates of thermal diffusivity and the melting point of rock to determine Earth's thermal gradient. Armed with these parameters, he calculated that the world was between 24 and 400 million years old, in line with calculations by the German polymath Hermann von Helmholtz.*

Turning his attention to the sun, Kelvin presumed it radiated energy due to gravitational collapse. From the rate at which it radiated heat, he estimated the age of the sun to be around 20 million years old, clashing with geological evidence that suggested a much older Earth. Kelvin wasn't overly sympathetic to the field of geology; some older practitioners remained bound to the doctrine that Earth was ageless, an assertion incompatible with the conservation of energy. Yet geologists were not monolithically ignorant of theoretical physics, and young geologists had rather more sophisticated views. Although they agreed that the world must have a finite age, they were every bit as meticulous with data as Kelvin was with theory. This led to a clash between two seminal figures in science. While Darwin's estimate for erosion of the Weald was only intended as a ballpark figure, Kelvin seized upon it as "absurd" and it was removed from later editions. Darwin bemoaned this trenchant dismissal from the finest scientist of his age as his "sorest trouble."

Still, evidence for an ancient Earth accumulated apace. Theory and evidence seemed diametrically opposed. Contradictions, however, tell us that something fundamental is askew, provided we know where to look. With Kelvin becoming increasingly strident, his friend and former assistant John Perry was approached for comment. Perry thought it unlikely that the calculation was the issue, noting: "I have sometimes been asked by friends interested in geology to criticize Lord Kelvin's calculation of the probable

---

* Polymath seems an understatement for Helmholtz. In physics, he undertook pioneering work on energy conservation, electrodynamics, and thermodynamics. In medicine, he advanced nerve physiology and psychological perception of sound and vision. In addition, he authored tomes on the philosophy of science and critiques of society. Some people seem to exist solely to make the rest of us feel like massive underachievers.

age of Earth. I have usually said that it is hopeless to expect that Lord Kelvin should have made an error in calculation." Instead of focusing on the numeric aspects, Perry reexamined Kelvin's assumptions. And there, he detected a leap in logic subtle enough to have evaded attention: If, instead of being molten, heat was transferred much more efficiently at the core, then Kelvin's estimate would be utterly confounded. Physicists already knew of just such a mechanism: convection, the primary method of heat transfer in fluids.

Factoring this in, Perry estimated the world to be at least 2 or 3 billion years old. Spotting Kelvin's error, he contacted his former patron privately to inform him of it. This was either misunderstood or ignored, forcing Perry to go public with his suggestion in *Nature* in 1895. For all its elegance, Kelvin's analysis was haunted by the specter of a rogue assumption, which rendered it undone. Perry's insight not only reconciled the geological evidence with mathematical physics, it also revealed something incredible: The core of Earth was an extraordinarily hot fluid. This was completely unexpected, arising organically as a logical consequence of data and theory. Today, we know that the planet's outer core consists of liquid iron and nickel, the motion of which gives rise to the earth's magnetic field.*

Less than a decade later, radioactivity was discovered, and shortly after this Einstein's theory of special relativity showed mass and energy to be equivalent. By 1920, British astronomer Arthur Eddington proposed that stars liberated vast quantities of energy by fusing small nuclei together, generating energy from mass. This is known today as nuclear fusion and is precisely how stars like our sun produce their energy. The discovery of this mechanism

---

* The evidence for this now is overwhelming, but Perry's 1895 papers were sorely underappreciated. Up until as late as the 1960s, there were geophysical models of Earth as a solid sphere! It is often claimed that Kelvin's ignorance of radioactivity threw his Earth-age estimate out. This isn't true. While it caused him to drastically underestimate the age of the sun, it didn't apply to his Earth-age calculations. Even if he had factored in the heat from radioactive decay, his result would have barely changed. Again, Perry deserves more credit for showing this than the history books have given him.

sounded the death knell for the young-sun theory, and the vestiges of young-Earth theory crumbled away as a new era of science began. Today, techniques made possible by these new discoveries—such as radiometric dating—allow us to date Earth accurately to about 4.54 billion years old.

Kelvin* was armed with the finest theoretical tools of his day and a mind capable of wielding them, yet his status as a pre-eminent scientist did not shield him from being incorrect. Nor did it impede the race to reconcile theory with observation. The elegant deductive reasoning underpinning the reconciliation not only resolved the contradiction but also gave rise to a completely new understanding of our planet. This is all the more staggering when we consider that the incredible finding about Earth's molten heart was not directly sought but emerged naturally from the synthesis of evidence and theory. This is precisely how science self-corrects; no matter how elegant or powerful an idea might be, if strong evidence contradicts it, it must be modified or jettisoned. A common criticism of science is that it is fickle, always changing its mind, seemingly on a whim. But this complaint misunderstands the scientific method—refining our ideas as new evidence emerges is not a bug but a feature.

It is a beautiful and deeply satisfying thing when theory and experiment agree, but an unexpected result might imply that there is something—perhaps something wondrous—waiting to be discovered. Sometimes discoveries are serendipitous affairs, with unexpected results paving the way for revolutionary conclusions. Indeed, many phenomena we rely on today arose from anomalous results. Henri Becquerel's discovery of radioactivity stemmed from an unexpected exposure of a photographic plate by a lump of uranium; Alexander Fleming's life-saving research on penicillin

---

* Some young-Earth creationists claim Kelvin as one of their own. They are completely mistaken; first, Kelvin questioned the time frame for evolution, not the principle itself. Second, Kelvin suggested more than 20 million years as an age for Earth, not a paltry few thousand. He also rejected chemical models of the sun precisely because they would only allow for a very young (10,000-year-old) Earth.

began when he returned from holiday to find an errant fungus had grown and killed off the staphylococci bacteria he'd plated; and the principle of microwave heating was first observed when engineer Percy Spencer ambled too close to a magnetron and found to his annoyance that it had melted the chocolate he'd bought. Curious experimental findings drive theory, and unique theoretical predictions point the way for experiment. These two aspects of science—experiment and theory—are equally vital and deeply intertwined.

But what determines if something is scientific? What line separates the science of astronomy from the superstition of astrology? After all, they both concern themselves with the motion of heavenly bodies. Why do we consider radiotherapy scientific and reiki pseudoscientific when "energy" is central to both practices? An intuitive feel of what science is and what it is not isn't enough if we wish to pin down the precise boundary between the reasoned and the ridiculous, the scientific and the specious.* One answer to that question was to stem from the most unlikely cradle.

Vienna in 1919 was a city in strife. The Great War had ended, but an Allied blockade remained. Food was in short supply and civil unrest palpable. Bavaria and Hungary had just been declared Soviet republics and Austrian communists plotted toward a central-European communist bloc. A coup was planned but, before it could be executed, Viennese authorities arrested the ringleaders.

After the communist uprising was foiled, socialists took to the streets in June, protesting the conditions in the city. Among them was Karl Popper, days shy of his 17th birthday, who aligned himself with the devoutly Marxist Social Democratic Workers' Party of Austria (SDWPA). During the protest, an attempt by communist elements to storm the jail and liberate their comrades

---

* In philosophy of science, this is known as the demarcation problem. For many scientists, such portentous arguments are both draining and of negligible utility. Richard Feynman is alleged to have remarked that "philosophy of science is as useful to scientists as ornithology is to birds." For all Feynman's brashness, however, he would have been the first to agree that distinguishing science from nonscience is a vital undertaking.

sparked a violent ruckus. In the chaos, police opened fire on the unarmed crowd, killing several protestors. The bloodshed left Popper distraught. The SWPDA response, however, was almost celebratory. This reaction stemmed from a sincere belief in the Marxist teaching that class war and revolution were harbingers of a communist future, and the dead were viewed as unavoidable casualties who heralded the inevitable revolution. But the carnage Popper witnessed threw these axioms into sharp relief and he grew steadily more uncomfortable with the tenets he had once implicitly accepted. On reflection, he found himself "shocked to have to admit to myself that not only had I accepted a complex theory somewhat uncritically, but I had also actually noticed quite a bit that was wrong."

Marx's doctrine of historical materialism bothered him especially. It insisted all human history was driven by entirely material considerations. Marx and his adherents called it science, yet Popper felt it sufficiently nebulous to explain away any fact brought before it—or to reframe slaughter as a sign of progress. While Popper remained a lifelong socialist, the realization that his pacifism and skepticism over Marx's pronouncements were not shared by his peers led him to renounce Marxism.

After this foray into political activism, Popper floated through a succession of careers, from construction to cabinet-making. At 24, he qualified as a schoolteacher, starting an after-school club for endangered children with Josefine Anna Henninger, whom he later married. By 1928, he had obtained a doctorate in psychology. As fascism rose across Europe in the early 1930s, the Poppers were forced to flee Vienna. In exile, Popper's thoughts focused on two much-lauded men whose ideas dominated conversations among European intellectuals of Vienna: Albert Einstein and Sigmund Freud. Einstein had made audacious predictions about the nature of reality, suggesting that space itself was warped by mass, resulting in the force we feel as gravity. His field equations quantified the consequences of this interpretation precisely, predicting that light would "bend" around an enormous object such as the sun due to its gravitational distortion, even calculating exactly how much. In 1919, Arthur Eddington and colleagues experimentally verified this prediction by observing the deflection of starlight around the sun during a solar eclipse, making Einstein a household name. Freud was a celebrated therapist to Vienna's upper-middle class, his esteem stemming from his status as the father of psychoanalysis. In his most significant work, *Die Traumdeutung* (*The Interpretation of Dreams*), Freud claimed that dreams were subconscious wish fulfillment.

As with Marxism, the label of science was applied to both men's work. Yet Einstein's ideas seemed exceedingly fragile, with clear predictions capable of being torn asunder. Despite this vulnerability, they passed every experimental hurdle. The same could not be said of Freud. A patient who dreamt of the mother-in-law she abhorred doubted Freud's wish-fulfillment paradigm. Freud countered by asserting her "true" wish was for him to be wrong—preserving his conjecture in spite of contrary evidence. Einstein's ideas made specific and testable predictions, whereas Freud's pronouncements were amorphous and capable of being massaged in such a way as to be interpreted as true after the fact. While Einstein's ideas had a testable vulnerability, Freud's were insulated from criticism. Popper's insight was to suggest *falsifiability* as that which separates the scientific from the pseudoscientific; if one could envision a single experiment whose results might contradict a hypothesis, then that idea was a scientific conjecture.

A *scientific hypothesis* must make specific predictions that can be tested. Without these, an idea cannot be considered scientific. Crucially, falsifiability does not mean a hypothesis is false—merely that it can in principle be disproven if it were. "It will rain in New York on Tuesday" is falsifiable; if it doesn't rain, the conjecture is dismissed. A psychic asserting that incorporeal spirits whisper to them isn't falsifiable, even though it is likely false. Scientific ideas are tested to destruction and if the weight of evidence contradicts a hypothesis, it is revised or dismissed. Strictly speaking, this means that no hypothesis is ever "proven." Instead, evidence consistent with a given hypothesis mounts up over time. Ideas that withstand critical scrutiny eventually become theories. These too demand revision when evidence contradicts them. Newton's law of motion seemed impervious to challenge for over 220 years, correctly predicting the movement of bodies from the minuscule to the celestial. In 1905, Einstein showed that Newton's laws don't hold at velocities close to the speed of light, which led to extraordinary refinement of our understanding of nature.

Falsifiability is fundamental to the scientific method.* It insists that scientists not only look for corroborating observations but also actively test their ideas with the utmost scrutiny. This explains why astrology is not science—its claims are so nebulous and vague they are impossible to test. Like Freud's psychoanalysis, astrological readings can be reinterpreted as correct after the fact. The science of astronomy, by contrast, makes highly specific and testable predictions. Reiki claims to tap into a curative universal energy but proffers no evidence or even definition for this central life force. Arguably, reiki is testable, but clinical investigations to date suggest no evidence for efficacy, rendering it directly falsified. Radiotherapy, on the other hand, is supported by volumes of data from theory and experiment.

Ideas that are not testable are not science and those that fail to withstand the trials of investigation should be dismissed. But with pseudoscience, believers often resort to special pleading and anecdote to explain away the weaknesses and failures of their convictions. Faith certainly plays a role. Evolution, for example, is science—its testable claims have withstood a barrage of experimentation. Creationism is not—with no testable predictions, it amounts to nothing more than a religious "just so" story. This

---

* Philosophical concepts are perpetually contentious; Popper's are no exception. There are voluminous tomes on the demarcation problem, but arguments tend to get obtuse quickly. I leave it to readers to dig deeper if they're so inclined.

is a vital difference between science and faith. In science, even a modicum of conflicting data may be enough to slay an idea, no matter how elegant. Faith—whether religious, political, or otherwise—demands that certain axioms be placed far beyond the realm of inquiry, and jettisoning evidence to preserve belief is often seen as a virtue.

It is, alas, not straightforward to distinguish science from pseudoscience. Many dubious ideas masquerade in the stolen robes of science, eager to lend an illusion of veracity to the vacuous. It is unfortunately all too easy to fall victim to dubious claims, much to our collective detriment. But, while picking apart science from nonscience is not an easy task, there are some vital things we can consider when confronted with anything purporting to be science. A non-exhaustive list might include:

- *Quality of evidence:* Scientific claims are underpinned by supporting data and clear description of the methodology used. If, however, a claim relies largely on anecdote and testimonial, it should be considered suspect.

- *Authority:* Scientific claims don't derive their authority by virtue of coming from scientists. A scientific claim's acceptance stems from the weight of the evidence behind it. By contrast, pseudoscientific claims often focus around ostensible experts or gurus rather than evidence.

- *Logic:* If an argument is presented, every link in the chain must connect, not just a few. Non sequiturs suggest dubious conclusions. Overly reductive claims that suggest single causes or cures for complex situations or conditions should also be treated skeptically.

- *Testable claims:* Falsifiability is paramount to gauging the validity of a claim. If it cannot be proven wrong, then it is not scientific. Similarly, science pivots on reproducibility. That which cannot be verified by independent investigation is likely to be pseudoscience.

- *Totality of evidence:* The hypothesis must consider all the evidence and not just cherry-pick only corroborating evidence. If the claim is consistent and compatible with all the evidence to date, then it is usually reasonable to accept it provisionally. If, however, it clashes with the weight of previous data, testable reasons for this disconnect must be suggested.
- *Occam's razor:* Does the claim rely on a multitude of supplementary assertions? If an alternative hypothesis better explains the available data, strong evidence would have to be provided to justify additional assumptions.
- *Burden of proof:* The onus is always on those making the claim to support it rather than for others to disprove it. Attempts to shift the burden of proof are a warning sign of bad science. Claims that pivot on special pleading to justify a lack of evidence (including claims of conspiracy) are hallmarks of pseudoscience. As a rule of thumb too, one should be mindful that claims made in the spirit of inquiry are more likely to be scientific than those made in the spirit of justification.

These items are no substitute for judicious reasoning, but they are questions we have to ask when faced with new claims. Understanding these basic precepts of science is profoundly useful, even when the issue at hand isn't a directly scientific one. Psychologist Thomas Gilovich puts it like this:

> Because science tries to stretch the limits of what is known, the scientist is constantly thrust against a barrier of ignorance. The more science one learns, the more one becomes aware of what is not known, and the provisional nature of much of what is. All of this contributes to a healthy skepticism toward claims about how things are or should be.

Distinguishing whether such claims have merit or not can, of course, feel like a Sisyphean task—we live in an era where pseudoscientific pronouncements frequently overwhelm us with an ocean of non-sense. In a storm of half-truths, it's understandable that many of us are driven to inertia. But to disengage is to sleepwalk into disaster; we cannot hope to face down the looming specter of climate change, nor the myriad other challenges ahead, if we are overtaken by apathy. Sagan's lament was percipient but not an inevitability. Learning to sift the scientific from the pseudoscientific is vital if we are to protect ourselves from charlatans and fools.

RISE OF THE CARGO CULT

Cargo-Cult Science, Health Fearmongering,
and the FBI's Shoddy Convictions

Richard Feynman's baroque life was defined by an insatiable curiosity. As a young physicist working on the Manhattan Project in Los Alamos, he took up safe-cracking to entertain himself through the long New Mexico nights, amusing himself by leaving cryptic notes behind. Predictably, the appearance of phantom notes inside an atomic-bomb facility created panic that a saboteur was loose, so Feynman took up bongo-playing instead. After the war, he became an eminent theoretical physicist, winning the 1965 Nobel Prize. But he is equally well known for his brilliant teaching, rooted in captivating stories. Perhaps one of his most enthralling tales concerns islanders in the South Pacific and their mysterious cargo cults.

These curious religions sprang up in the wake of the Second World War, peppered in isolated outposts throughout the Pacific Ocean where indigenous Melanesian islanders had, for generations, lived blissfully unaware of the technological developments of the wider world. Hostilities among warring nations brought the front line–and all its trappings–to their doorstep. The Japanese arrived first, and with them came airlifts of medicines, food, and equipment. Soon after, Allied forces set up military bases on the islands, with a steady flow of supplies the likes of which the natives had never seen. Some of the soldiers even shared supplies with bemused natives. When the war came to an end, the soldiers and their cargo abruptly departed as quickly as they had appeared.

To some native islanders, the servicemen and their cargo achieved religious dimensions. The writer Arthur C. Clarke once observed that "any sufficiently advanced technology is indistinguishable from magic," and as religion itself tends to demand selective transgression from observed laws of nature, it's hardly surprising that many islanders interpreted the cargo drops as otherworldly. This reverence was exploited by charismatic "big men" who formed cults amid the abandoned structures. As with all religious sects, beliefs varied with geography and prophet, but all conformed to the idea that the cargo was of divine providence. On the fringes of Melanesian society, cult leaders preached that the faithful would be rewarded with jeeps, food, and clothing.

To coax the ancestral deities who they believed control this, the islanders adopted behaviors they had observed, transforming them into highly stylized rituals. They constructed elaborate replicas of what they had seen: expansive grass runways and model control towers, with artistic embellishments suggestive of the mysterious communications and military equipment they had seen the servicemen use.

On the island of Tanna, believers waited for the return of a deity they called John Frum, who, it was promised, would return to the faithful with cargo. His appearance varied with retelling—sometimes white, other times black. Often, he took the appearance of a Second World War-era serviceman. When documentary maker David Attenborough asked believers what Frum looked like, he was told: "'E look like you. 'E got white face. 'E tall man. 'E live 'long South America." Yet, despite their devotion to relic and ritual, no cargo was ever forthcoming—the runways remained desolate, the towers silent.

Believers attributed this failure to the whims of Frum. But the more fundamental problem was that the Islanders had merely replicated the aesthetics of the technology they'd seen, with no understanding of the underlying ideas behind them. By way of analogy, Feynman coined the term *cargo-cult science* to describe phenomena that ape the theater of science yet are completely insulated from the realities of the scientific method. He noted that meaningful science must be underpinned by "scientific integrity, a principle of scientific thought that corresponds to a kind of utter honesty—a kind of leaning over backward. For example, if you're doing an experiment, you should report everything that you think might make it invalid—not only what you think is right about it."

Feynman's concern was well placed: a semblance of science can mask extremely dubious currents. The very aura of "science"—justified or not—brings with it an implicit ring of authority, making us far more likely to accept it uncritically. Unfortunately, a superficial veneer is too often the Trojan horse that spirits in all manner of falsehoods. Suspect claims are too often draped in the borrowed robes of science. And today, with the explosion of information available to us, this facet makes separating junk science from valuable insight an arduous task.

If there's a unifying theme to all we've touched upon so far, it might be the observation that we are a species riddled with contradictions. These contradictions extend to science too; even those who dismiss science still tend to cling to it if they think it supports their argument. Academic journals of homeopathy exist, their articles extolling the virtues of ritualistically shaken water, although the tenets of homeopathy crumble under the most rudimentary analysis. Such journals are devoted to a fiction, yet in all appearance they resemble genuine academic writing, with neat citations and complex jargon. However, this superficial resemblance to science is nullified by the fact that the proponents of homeopathy insist on rejecting the overwhelming evidence against their beliefs. This means these journals fail utterly at the basic integrity the scientific method requires. These journals embody Feynman's cargo cult, constructing an impotent effigy of science that is no more genuine than the grass runways of the Pacific Islanders.

Part of the reason for this is that we live in a world transformed by science and medicine, much of which is hard to understand or unintuitive. Yet as the success of science has been undeniable, there is an implicit acceptance, even if grudgingly granted, of science as authoritative. Cargo-cult scientists have never been more plentiful, thanks in part to the democratizing nature of the internet—while access to the wealth of the world's information is available quite literally at our fingertips, the downside is that nonsense and misinformation perpetuate with the same virulent rapidity. The sheer volume of information available means that sifting signal from noise is not always an easy task, and in this environment cargo-cult scientists thrive, emulating the apparent authority and respectability of science to advance dubious beliefs. Meticulously referenced websites and forums exist on all topics imaginable, from New World Order conspiracy theorists to aromatherapy, presenting outlandish ideas in a distinctly scientific style, despite the theories propounded being totally devoid of scientific merit.

Such cloaking in the clothes of science can, unsurprisingly, fool the unwary into thinking that these claims are legitimate, especially if the claims are sufficiently worrying or presented with enough aplomb. Aspartame, a common sweetener used in diet drinks, was the subject of a 1995 email scare. In the widely circulated warning, one Dr. Nancy Markle reported ominous findings from a scientific conference on the ill effects of the sweetener, with side effects including lupus, cancer, Alzheimer's, and even Gulf War Syndrome. The warning was written in a semi-formal style, with scientific-sounding terminology and patchy referencing. Recipients of the email—presumably unaware of the danger of taking health advice from spam emails—circulated it to friends and colleagues. Concerned recipients were completely hoodwinked by the illusion of science on display, and even reassurances by public-health bodies that the claims made in the email were bogus did nothing to abate the flurry of sharing. Nor did the revelation that Dr. Markle was in fact a fictitious creation of notorious health-crank Betty Martini serve to stem the rising hysteria. Today, the myth is ingrained in the public consciousness to a seemingly indelible degree. In 2015, Pepsi removed aspartame from its diet range after conceding to fears the sweetener would impact sales, before reintroducing it a year later after customers objected to the new taste. Sadly, the message seems to be that the numerous studies debunking every aspect of the myth are readily drowned out by the sheer volume of cargo cultists incapable of distinguishing real science from imitation.*

---

* You'll not be surprised to learn that the infamous Joseph Mercola's empire of pseudoscience stands firmly behind the myth, referring to aspartame as "the Most Dangerous Substance on the Market that Is Added to Foods."

The internet has also revitalized old myths, and the appropriation of scientific presentation has made them exponentially more difficult to spot. Fluoride conspiracy theories are one such example. These are nothing new; in the classic Stanley Kubrick film *Dr. Strangelove,* the unhinged General Ripper was an adamant believer that water fluoridation was a communist plot that had rendered him impotent—a send-up of the more outlandish claims of anti-fluoridation propaganda and cold-war paranoia. Even by the time of the movie's release in 1964, such conspiratorial sentiment was already well worn; in the intervening decades, numerous academic studies have reaffirmed water fluoridation as a safe and effective method of reducing dental cavities, which in principle should have confined this movement to nothing more than a historical footnote. Instead the internet age has seen something of resurgence in the anti-fluoridation movement. Worryingly, much of this is buttressed by scientific-sounding websites listing all sorts of odious side effects, from cancer to depression. While these fearmongering sources are bereft of scientific merit, this does not seem to be an impediment to their acceptance.

I have experienced this first-hand. In 2013, a number of Irish opposition politicians brought motions to the national parliament, Dáil Éireann, demanding the removal of fluoride from water supplies. Their motions quoted scientific-sounding reports warning of a plethora of ominous consequences arising from fluoride—Alzheimer's, Down's syndrome, depressive illness, diabetes, and cancer, to name but a handful. A little investigation showed that they were being fed these dire claims by self-appointed experts from the anti-fluoride fringe. These reports were archetypical cargo-cult science, filled with pseudoscientific prose, serious errors, and contorted logic. Beyond the veneer of scientific inquiry, the reports bore precious little adherence to the scientific method. Yet, despite the vapidity of the work, a worrying number of journalists and politicians were unable to distinguish this from an abundance of genuine scientific works countering such hyperbolic claims.

In response, I authored pieces for *The Irish Times* and *The Guardian*, outlining why these reports were flawed. I also debated with the politicians leading the charge, explaining why these frightening claims lacked merit. This made me something of a hate figure for the fringe groups, and rather predictably I was lobbed with assaults on everything from my motivations to my scientific aptitude. While the motion was eventually defeated in the Dáil, movements to ban fluoride still crop up with depressing regularity not only in Ireland but also in other nations where fluoridation is part of public-health policy, including the US, Canada, New Zealand, and Australia, supported by nonsense masquerading as science. Rather frustratingly, whether born of genuine scientific ignorance or a cynical attempt to garner votes from the disaffected, there does not appear to be a shortage of local politicians ready to take pseudoscience at face value.*

Cargo-cult science is often simply a means to sell a ridiculous idea, basked in the reflected glow of scientific validity. Few avenues of science are as misunderstood or as shamelessly abused as quantum mechanics (QM). This fascinating branch of physics deals with the behavior of tiny subatomic particles and is very different from the macroscopic world we inhabit. The domain of the very small can be wholly counterintuitive; entities in the quantum realm shed the classical dichotomy between particles and waves and instead embrace properties of both, a concept known as wave–particle duality. While far beyond our scope in this book, QM also yields several exotic and esoteric ideas. Quantum entanglement is one example, where particle states are intrinsically linked even when separated by great distances. Quantum tunneling, where particles "tunnel" through a classically forbidden barrier, is another. In some

---

* I spoke with several instigating politicians, as I was perplexed that they could be taken in when comprehensive myth-busting was on hand from the Irish Expert Board on Fluoride and Health—itself set up in response to the last emergence of anti-fluoride panic a mere decade before. Ineptitude only goes so far in explaining it; I suspect the majority of the political backers of the bill were merely harnessing the political capital of the vocal anti-establishment movement. Their willingness to jeopardize public health to do so spoke volumes about their priorities.

interpretations, the very act of observing a process can affect its outcome. These phenomena are fascinating, far removed from our innate understanding of the world. Quantum physics raises profound scientific and philosophical questions about the nature of reality and the limits of our understanding.

Frustratingly, there exists a healthy market for a mangling of the lofty ideas behind QM that pilfers the terminology. Quantum mysticism is a new-age belief that reduces QM to a universal deus ex machina for any old nonsense, adding an illusion of depth to patently vacuous philosophies. These tend to be haphazard platitudes that tout QM terminology with no real understanding of the phenomena in question, served with a poorly digested pastiche of Eastern philosophy. They so exasperated physicist Murray Gell-Mann that he labeled them "Quantum flapdoodle." The infamous Deepak Chopra is no stranger to abusing the nomenclature of QM to push dubious—yet profitable—philosophies, with bestselling books such as *Quantum Healing* to his credit.

What makes this abuse of science especially galling is that even a cursory understanding of QM reveals that it applies chiefly to the realm of the incredibly tiny, smaller even than atoms. The astute reader will no doubt notice that human beings are considerably larger than a quantum particle, and ham-fisted attempts to apply QM to human affairs tend to miss the point by a wide margin.

Quantum quackery is, of course, a transparently outlandish attempt to imbue some new-age beliefs with a forged stamp of scientific authenticity, but cargo-cult science also manifests in less obvious and more damaging ways. As Feynman noted, science requires above all else a devotion to integrity, and with it a willingness to profess the weaknesses and limitations of our measurements and theories. This isn't mere idealism or self-flagellation on behalf of scientists; true science requires an unfailing willingness to countenance an alternative hypothesis that might better describe the phenomenon in question. In order to do this, we require a quantifiable grasp of error, an understanding of the limits of certainty and, crucially, a willingness to question our own theories and results objectively. This can jar with our deep craving for certainty. To banish the demon of uncertainty, we often reject the possibility of error or alternative explanation. In *The Hitchhiker's Guide to the Galaxy*, Douglas Adams parodied this flawed fixation by way of a disclaimer against the guide's multitudinous errors: "The Guide is definitive. Reality is frequently inaccurate." The self-insulating thinking that Adams parodied is unfortunately common, tantamount to substituting an almost-religious faith in the place of the logical deductions demanded by the scientific method.

When those who should know better engage in cargo-cult science, the consequences can be tragic. In February 1981, a young woman in Washington, DC, was the victim of a horrifying ordeal at the hands of a gun-wielding assailant. His attack was brutal; he raped and sodomized her, leaving her bound and gagged after taking whatever valuables he could. Through the entire assault, she got only a single glance at her abuser, through the dim illumination from a street lamp creeping through the curtain—enough only to know that her assailant was a young, clean-shaven Black man. A few weeks after this, 18-year-old Kirk Odom was stopped by a police officer on a totally unrelated matter. The questioning officer produced a sketch of the assailant, asking Odom whether he thought there was some resemblance between the sketch and himself. Odom disagreed, and after taking his details the officer waved him on. Days later, police showed up at the address Odom had provided, arresting him for the brutal rape and robbery. The case against Odom was incredibly flimsy; the composite drawing the victim had provided described a Black man of medium complexion, whereas Odom himself was dark. The victim had tentatively identified Odom in a photograph, but this was hardly proof positive. Being young and clean-shaven, Odom's mugshot already stood out among the sea of scruffy middle-aged faces. This evidence was equivocal at best; eyewitness identification is hardly sound evidence, as we've seen already. Besides, Odom had an excellent and well-corroborated alibi: On the day of the incident, he had been with his family to celebrate the birth of his sister's daughter. Odom quite reasonably expected charges to be dropped and his nightmare to end.

However, despite the paucity of evidence against him, the prosecution had a trump card: a single hair found at the crime scene. In court, special agent Myron T. Scholberg took the stand. Affiliated with the FBI's microscopic analysis unit, he introduced himself as a world leader in the science of hair microscopy. Scholberg's testimony sealed Odom's fate; the hair at the crime scene was an exact match, completely indistinguishable from Odom's hair. Were this not damning enough, Scholberg stated this level of match was a "very rare phenomenon," one he had only encountered a handful of times in all his years of expertise. To the jury, infallible science had spoken; Odom's protestations of innocence were dismissed as utterly untenable.

For 22 long years, Odom languished in prison. Even after release, he spent another nine on parole, a registered sex offender. Lacing insult with injury, this also put his relationship with his daughter—who was only an infant when he was locked away—under huge strain. It is impossible to fathom how keenly this blow must have landed, yet Odom's protestations of innocence were roundly ignored. The infallible might of forensic science sealed his fate, and an explosion of popular-culture shows extolling the virtues of crime-scene investigation reinforced the perception that such evidence was impossible to challenge—a phenomenon dubbed the CSI effect. But in December 2009, in a separate case, the Public Defender Service (PDS) for the District of Columbia won the exoneration of Donald Gates. Gates had served 28 years in prison for a rape and murder he did not commit. Key to his exoneration had been the demolition of the seemingly watertight forensic analysis—including microscopic hair analysis extraordinarily similar to that which had convicted Odom.

The overturning of Gates's verdict caught the attention of Sandra Levick, Odom's original public defender. In the intervening years,

Levick had risen to chief of the PDS's Special Litigation Division. She had not forgotten the shoddy evidence against her former client. In February 2011–30 years after the original crime–Levick filed a motion for new DNA testing under the District of Columbia Innocence Protection Act. From long-locked evidence boxes, the fragments of the crime scene emerged; stained bed sheets, a bathrobe, and the telltale hair. With these relics of the original crime, Levick ordered retesting with modern techniques. They told a galling story: the semen at the crime scene was not Odom's, instead matching that of a convicted sex offender. Mitochondrial testing of the hair excluded Odom as a suspect. His entire conviction had been a farce. Levick filed a motion to vacate Odom's convictions on grounds of his innocence in March 2012, and Odom was exonerated on 13 July 2012–his 50th birthday–after spending most of his life in prison for a crime he had not committed. This vindication of innocence triggered a surge of interest in historic cases and raised an unsettling question: Just what had gone wrong with the microscopic hair analysis?

It's easy to see why juries were fooled by the illusion of scientific rigor. The 1977 FBI handbook on the subject was peppered with technical terms and, at its zenith, hair analysis boasted 11 full-time agents, 2,000 cases, and over 250 court appearances a year. Yet the flaws were apparent to those who knew where to look. At an internal 1985 FBI conference in Virginia, concerns were voiced by the chief scientific officer of the London Metropolitan Police, who pointed to the "reluctance among examiners in the United Kingdom to examine hairs because of the generally low to very low evidential value put on most hair matches by the average hair examiner in the UK." At the same conference, New York criminologist Peter De Forest referenced the Odom case, slamming FBI conclusions as "very misleading" and "not substantiated by any data." In spite of this, senior FBI laboratory member Harold Deadman (an apt name for someone in criminal forensics) insisted the FBI were "believers in hair comparisons."

Belief, however, was the problem. Refusal to look at the method's weakness objectively was the hallmark of a cargo-cult mentality that had infected the FBI. Some were not totally oblivious to the tragedy waiting to happen. Agent Fred Whitehurst raised the alarm on the dubious nature of the technique on numerous occasions, penning 237 letters to his superiors on the subject between 1992 and 1997. They were not heeded because hair analysis "got convictions." Ignored by the FBI and outraged over the pseudoscience at play, Whitehurst eventually became a whistle-blower, alerting determined public defenders to the problems with the method. This, coupled with improvements in scientific DNA testing, led to a slew of overturned convictions.

A 2009 report by the National Academy of Sciences delivered a damning insight into what had transpired; the microscopic analysis championed by the FBI, despite being cloaked in scientific-sounding terminology, was profoundly lacking in basic scientific integrity. To compare two samples and determine whether they are a match requires some solid statistics on the distribution of types in the general population. As these didn't exist, the FBI expert testimony stating a match was a "very rare phenomenon" was a complete fiction. This cargo-cult science had all the theatrics of the real thing, but none of the substance; the method had the veneer of forensic science, but was just a guessing game that sent innocent people to prison.

It's also crucial to note that systematic issues of race and racism may have reared their ugly head, not only in Odom's case, but in many others. Identifying suspects of different race is fraught with pitfalls, such as cross-racial identification bias: the diminished ability to clearly identify features and faces in people from different racial groups. This might not be motivated by any hatred, but renders much eyewitness testimony suspect. Of the 375 people exonerated by the Innocence Project through DNA evidence, 109 involved cross-racial misidentification. More alarming still, 60 percent of all those exonerated were African American, a staggeringly high proportion of the total. The racial disparity in America's penal system is especially striking, with Black offenders garnering substantially more prison time for the same crimes than white offenders, even adjusting for previous offenses. While far beyond the scope of this book, these findings seem indicative of a much deeper structural problem with assumptions about race and guilt than many are willing to concede.

The month of Odom's exoneration, the Department of Justice and the FBI announced a joint review of convictions that pivoted on analysis of hair evidence, assisted by the Innocence Project and the National Association of Criminal Defense Lawyers. Their report, released in April 2015, is damning, acknowledging that fatally flawed testimony was given in the vast majority of the trials involving microscopic hair analysis, including 32 cases that sent defendants to death row; nine of these people had already been executed by the time the report saw the light of day.

The problem was that the FBI relied on a pseudoscientific technique, sold as legitimate forensic analysis. Just as in the Sally Clarke case, the defense team and jury were bamboozled by the stamp of science that the FBI had falsely affixed to the method, and, blinded by this, they failed to ask the most pertinent of questions. At the time of writing, several convictions have been overturned and thousands more questionable cases are to be analyzed. Yet despite this, when confronted with insurmountable evidence that their tests were less than perfect, many adherents displayed an almost-religious cargo-cult certainty that this could not be the case. Exasperated extracts from the 2009 NAS report allude to this very problem and the wider issues of discounting the possibility of error or uncertainty in forensic analysis:

> Some members of the forensic science community will not concede that there could be less-than-perfect accuracy either in given laboratories or in specific disciplines, and experts testified to the committee that disagreement remains regarding even what

constitutes an error . . . The insistence by some forensic practitioners that their disciplines employ methodologies that have perfect
accuracy and produce no errors has hampered efforts to evaluate
the usefulness of the forensic science disciplines.

Cases like this remind us that it is supremely important to distinguish
between genuine scientific inquiry and the theater of investigation.
While it is positive to see that science is a trusted enterprise by the
general public, good science requires constant questioning to prevent
it from becoming a mere illusion or, worse again, an argument from
authority–the polar opposite of scientific integrity. All scientific
claims need to be transparent and critically examined; this is the
motivation for peer review, the process where scientists submit their
work and data to be ruthlessly evaluated by other generally anonymous and unaffiliated researchers. These reviewers are tasked with
critically evaluating the manuscript at hand, finding weaknesses,
mistakes, or logical gaps that might invalidate the conclusions of a
body of work. This can be a frustrating and imperfect process, yet
this harsh devil's advocate approach is required to circumvent all
manner of mistakes and nonsense from perpetuating.

This strenuous testing of ideas and claims is a vital component
of science, but it is conspicuously absent from cargo-cult offerings.
Cargo-cult applications may well grasp the aesthetic of scientific
inquiry but not the integrity it demands, and their offerings are no
more effective than the grass control towers of the Pacific Islanders.
But while this is easy to state, the sheer volume of information we're
subjected to makes it incredibly difficult to distinguish the genuine
from the devious. How we discern the two is not immediately obvious, but in an age of perpetual information bombardment, sifting
the real from the illusory has never been more vital. To do this, we
need more than anything to hone skepticism and analytical thinking.
These are the most powerful tools we have to uncover the truth of
claims we're accosted with. It doesn't matter whether these claims
are scientific, political, or otherwise, the same methods can sift the

signal from the noise. Skepticism is simply invaluable if we're not to be manipulated or misled.

As an aside, there are small enclaves where modern-day cargo cults of the anthropological kind still thrive. In Tanna itself, members of the Yaohnanen tribe deify the most unlikely of candidates: Prince Philip of England, consort to Queen Elizabeth II.* According to ancient Yaohnanen legend, the son of the mountain spirit roamed far across the ocean and married a great and powerful lady. Prophecy foretold that in time he would return to them. And, as legends are wont to do, it acquired embellishment with each successive telling. Some said that the mountain spirit's son was the brother of John Frum. The tribe's people were aware of the respect that colonial officials afforded their queen, concluding that she was the long-heralded wife of the mountain spirit's son. This reasoning meant that Prince Philip was the living embodiment of the mountain spirit's offspring, and the royal visit to Vanuatu in 1974 cemented this view. After being informed of his reverence as a god, Prince Philip sent the villagers a signed official photograph. The villagers in return sent him a *nal-nal*, a traditional club for bludgeoning pigs. Whether he ever used it for its intended purpose we can only speculate.

---

* The elevation of Prince Philip to god is extremely amusing to those familiar with his reputation as a self-described "cantankerous old sod" notorious for spouting remarks that can be construed as blunt, inadvertently funny, or racist—sometimes a combination of these.

A HEALTHY SKEPTICISM

Cancer Cure Conspiracies, Chernobyl Fictions,
and How Scientific Skepticism Can Save Us

In December 2012, a child was abducted in England. The missing boy was seven-year-old Neon Roberts, who had had recently undergone treatment for a brain tumor (medulloblastoma) and urgently needed radiotherapy to save his life. The perpetrator of his abduction was not some hardened criminal or deviant reprobate, but his mother, Sally Roberts. Sally fundamentally disagreed with the concept of radiotherapy, refusing to take him to vital appointments. Time was of the essence, and concerns grew that delays were jeopardizing his life. With his mother adamantly opposed to the treatment, legal action ensued. The family courts quickly ruled that the life-saving radiotherapy was mandatory, but this was not something Sally could accept. In the run-up to Christmas, she absconded with Neon, sparking an intense manhunt.

Police officers countrywide scrambled to find the pair, eventually locating them in Sussex four days later. Back in court, medical opinion was unanimous: Without urgent treatment it was "highly, highly, likely he would die over a relatively short period of time." But Sally insisted that alternative medicine would cure her son, citing conversations with alternative practitioners. They claimed to be able to cure cancer with everything from diet to hyperbaric oxygen. The exasperated council for the hospital trust countered that these individuals provided no evidence of efficacy for their treatments. They hadn't even spelled medulloblastoma correctly, seemingly culling their descriptions of the illness from internet searches rather than from medical sources. To the shock of many, Sally dismissed her legal team, demanding time to find alternative medicines for Neon. Although sympathetic, however, the judge reiterated that time was not a luxury that could be afforded. Radiotherapy went ahead by ruling, despite a last-minute appeal by Sally.

Neon went on to make a full recovery. After the ordeal, Sally insisted: "Death by doctor is very common, but thankfully, because of the internet these days a number of us have educated ourselves . . . there's so many other options that we've been deprived of, denied."

There are few conditions with as terrifying a resonance in the public mind as cancer. It's an ominous reminder of our mortality and our deepest fear, and discussions about it are often avoided or couched in euphemisms. Over our lifetime, roughly half of us will be touched by the ubiquitous illness. The concepts articulated by Ms. Roberts are not unique; the belief that there are cures for cancer suppressed by medical science and the pharmaceutical industry is an enduring one. The Alliance for Natural Health* laments that "maverick cancer treatments are suppressed by the mainstream," while naturalnews.com proclaims: "The cancer industry worldwide is estimated at a 200 billion dollar a year industry. There are many in various associated positions within that industry who would be without a job if that cash flow dried up suddenly with the news that there are cheaper, less harmful and more efficacious remedies available. Big Pharmacy would virtually vanish."

These are not fringe beliefs; an estimated 37 percent of Americans believe that the FDA is so beholden to drug companies that they are suppressing "natural cures" for cancer. Robert Blaskiewicz, an authority on conspiratorial belief, defines the conspiratorial notion of "Big Pharma" as "an abstract entity comprised of corporations, regulators, NGOs, politicians, and often physicians, all with a finger in the trillion-dollar prescription pharmaceutical pie." It's no secret that pharmaceutical companies can both rake in staggering profit and engage in reprehensible

---

* The same outfit once ran a feature that attempted to cast aspersions on me, which included the line "he's young, he's hip, he's got cool hair." I imagine the intention was to ridicule me, but honestly this is the kind of thing I get printed on business cards.

behavior. To take but one example, GlaxoSmithKline was forced to pay a record-breaking $3 billion settlement in 2012 for a mixture of criminal and civil violations, including failure to disclose safety data and paying kickbacks to physicians to promote their drugs. This jibes with universal reservations about the pharmaceutical industry. Drug companies have ample cash at their disposal and a record of poor behavior with which physicians have often been complicit. Were a naturally occurring agent discovered that cured cancer and imperiled profits, it might seem plausible that unethical companies would suppress this.

Conspiracy conjecture is integral to many beliefs; devotees of alternative medicine insist that the pharmaceutical industry exerts sinister influence to quell proof of alt-med's efficacy, which we've seen previously with claims about cannabis. A quick internet search yields an abundance of alleged cures with the same DNA, including ketogenic diets, juicing, and even bleach. A related assertion is that cancer is a modern, man-made disease, designed to keep people sick and the medical industry profitable. Gurus frequently dismiss chemotherapy and radiotherapy as poisons, urging patients to opt for alternatives. Joseph Mercola, for example, casts aspersions on both conventional therapy and the pharmaceutical industry: "The cancer epidemic is a dream for Big Pharma, and their campaigns to silence cancer cures have been fierce."*

Details on this natural cure vary; sometimes it's a miracle diet, or alternative therapy, or common herb or plant. No matter the specifics, these claims are crafted of the same clay. Believers insist that drug companies cannot patent the agent in question, so they dedicate themselves to burying its unbelievable efficacy, aided by a complicit medical establishment. The narrative is seductive, clean, and seemingly explains everything. But the mere fact that a claim resonates with preconception doesn't mean it's correct. Let's examine the premises:

---

* Mercola's website gets much more traffic than reliable scientific sources such as the National Institute of Cancer.

- *Premise 1—There is a cure for cancer:* The immediate problem is that such a postulate betrays a concerning ignorance of cancer. Cancer isn't a single disease—it's an entire family of malignancies caused by unregulated division of mutant cells. It can arise from practically any type of cell, meaning prognosis and treatments vary hugely. Adding to the complexity, each cancer is unique to every patient precisely because it arises from mutations in their own cells. Consequently, the idea of a single magic bullet to treat all these forms with different causes and responses is extremely far-fetched. The premise is overly reductive, simplistic to the point of vapidity. This suggests the fallacy of the single cause at play, applied to both the illness and its treatment.
- *Premise 2—A cure would not be profitable:* Why would any drug company in their right mind suppress a cure for cancer? The canard about natural products being unpatentable is a convenient fiction—"natural" origin is no barrier to commercialization. Many of our drugs today are derived from compounds found in plants, herbs, and animals. The trick is to identify and synthesize active agents so that the dose is controllable. If turmeric or vitamin D cured cancer, pharmaceutical companies would be racing each other to isolate active compounds and proving their efficacy, not engaging in a long-running conspiracy. A universal cure for cancer would make its discoverers insanely rich, earning the discoverer fame, Nobel Prizes, and the world's eternal gratitude.
- *Premise 3—Cancer rates are manipulated by Big Pharma:* Cancer isn't a new disease—it has stalked humankind since our earliest days. There are 3,000-year-old Egyptian mummies showing evidence of the illness. By 400 BCE, Hippocrates had distinguished between benign and malignant neoplasms. Physicians of antiquity—unfamiliar with dissection—likened the protruding tumors they observed to crab

legs, the origin of the term "cancer." As we've seen, rising cancer rates are an artifact of improved societal health. It is a complete non sequitur to presume this implies sinister machinations. We're simply living longer, not succumbing in waves to plagues like cholera or smallpox.

On reflection, the central tenets of the grand conspiracy narrative fail to withstand even cursory probing. True believers will, of course, claim that all our information is wrong and manipulated by Big Pharma. But Occam's razor cuts deeply; accepting a grand conspiracy means believing millions of actors worldwide have worked in concert for decades, unconcerned about saving loved ones or themselves. Every drug company and regulatory body from the minuscule to the multinational would have to be involved, complicit in this scam against humankind, willing to forsake the profits and accolades a cure would bring. This is not only ludicrous, it's demonstrably unsustainable.* In reality, medical research involves doctors, scientists, and regulatory agencies, with differing roles and incentives. The narrative relies on collapsing this complex ecosystem into a monolithic "they," a conniving phantom upon whom blame for anything may be laid. The alternative hypothesis, by contrast, requires much less choreography; the grand conspiracy simply doesn't exist.

Why then does the story endure? A cynical answer is that it is invaluable for those pushing alternative cures, supplements, seminars, diets, and "wellness." A cry of conspiracy is special pleading, allowing charlatans or fools to dismiss findings contrary to their assertions, waving away the lack of evidence for their wares. To others, it holds allure by virtue of yielding easy answers to complex questions. Psychological research suggests that belief in conspiracy theory is intricately linked to a need for control.

---

* In my work on conspiracy theories, I've looked at the viability of drug companies and scientists "covering up" a cure for cancer. As you might expect, the mathematics suggested that, even if all involved were wholly unethical, the entire operation would be rapidly doomed to failure.

An illusion of control over an uncertain world is reassuring—if a person believes in a suppressed cure, they feel "protected" by this special knowledge. That a disease feared by so many is a locus for conspiratorial mumblings is hardly surprising. But these conspiratorial narratives pit patients against doctors. This fostered distrust is exacerbated by alternative gurus, who dismiss conventional treatment as a racket. The consequences are tragic; a 2018 study found that patients invested in alternative medicine were much more likely to refuse or delay effective cancer therapy, rendering them twice as likely to die within five years of diagnosis.

This grim statistic* highlights something important: Our very lives can pivot on our ability to evaluate the claims to which we're subjected. Cancer conspiracy may be an extreme example, but we are accosted each day by a cacophony of misinformation. Confounding things further, the greatest myths spring from seeds of truth, contorted into warped conclusions. Take dihydrogen monoxide (DHMO), for instance, a colorless, tasteless hydric acid found in acid rain, nuclear waste, and even human cancers. It causes major environmental damage and erosion and kills more than 360,000 people each year. It is potent enough to corrode metal. And yet, it is frequently found in our food supply and environment in abundant concentrations. Consequently, there have been numerous petitions to ban it in city chambers and parliaments the world over, from California to New Zealand. One Finnish survey, conducted in 2011, found that 49 percent of all respondents were in favor of restricting DHMO. Yet all this legislative zeal is somewhat misguided, for dihydrogen monoxide (H_2O) is usually referred to by its far more common name: water.

---

* Gains in life expectancy and quality of life are at least partially due to the huge investments in research undertaken by pharmaceutical companies, for all their vices. It is vastly expensive to research and discover new drugs, and much of the profit from drug companies goes into research. This doesn't absolve them of incidences of wrongdoing, but it paints a more complex picture than the villain archetype.

None of the negative aspects of water listed are false, of course. Rather, they're selectively curated, stripped of context. The DHMO parody exposes the problems that emerge not only from lack of scientific literacy, but from the exaggeration of carefully selected information. The same cherry-picking of facts is often employed for more sinister reasons. The claim that climate has always changed is frequently voiced by climate-change denialists, with the implication that global warming is overstated. But the fact that the climate has forever been in flux isn't contentious at all; what is alarming is the current rate—hugely in excess of anything natural—at which this change is occurring. This rate matters—there's a huge difference between bringing a car to a halt by the gentle application of the brake and running that same car full tilt into a brick wall. By presenting facts in isolation and devoid of context, one can be led to an impression completely at odds with reality.

We've seen before how climate-change denial is strongly ideologically motivated. The opposite end of the political spectrum is not immune from similar folly. Reducing our carbon footprint is imperative, and energy production accounts for the lion's share of greenhouse-gas emissions. Unlike heavily polluting fossil fuels, nuclear power has zero carbon emissions and is extremely energy efficient. The Intergovernmental Panel on Climate Change (IPCC) stresses that it is a vital part of mitigating climate devastation, with some estimates projecting a doubling in nuclear capacity required to stave off the worst ravages of climate change. Yet, despite the virtues of nuclear power, many green organizations adamantly oppose it with a single-word rebuke: Chernobyl.

In the early hours of April 26, 1986, a doomed experiment took place in the Ukraine that would etch the name indelibly on the world's collective consciousness. A perfect storm of ineptitude, obsolescence, and disregard for safety led to two colossal steam explosions, with enough kick to blow the 2,000-ton reactor casting clean through the roof of the reactor building. Exposed to air, the superheated graphite moderator burst into flames, showering the site with potent fallout.

The Soviet response was an unmitigated disaster. The helicopter tasked with dumping 5,500 tons of sand and neutron-absorbing boron to quench the fire collided with a crane, spiraling to the ground. Unprotected firefighters battled the inferno, unaware of the danger. No effort was made to stop hazardous material from contaminating the food chain, chiefly radioiodine 131. This radio-isotope has a half-life of merely eight days, but if ingested it accumulates in the thyroid. With basic precautions, health effects were easily avoidable; instead authorities insisted nothing was amiss, allowing locals to eat contaminated produce. This miasma of Soviet-era denial might have continued indefinitely had traces of fallout not been detected at a Swedish nuclear facility the next day, revealing the scale of the problem to the world. Finally, a full 36 hours after the event, the order to evacuate was given, but thousands had been needlessly exposed.

The world's worst nuclear disaster has been a linchpin of anti-nuclear campaigning for decades, heralded as proof positive than nuclear energy is inherently devastating. It is a name synonymous with death on a massive scale. Greenpeace asserts that the accident claimed 93,000 lives. Haunting images of childhood deformities and astronomical cancer rates in the region are burned into our minds. Chernobyl remains a byword for mass death.

But this perception doesn't map neatly onto reality. In the wake of the disaster, the UN Scientific Committee on the Effects of Atomic Radiation (UNSCEAR), the WHO, and others convened to monitor the health impact on people exposed. In 2006, after two decades of observation, the Chernobyl Forum reported that, of the firefighters exposed to huge doses and toxic smoke, 28 died from acute radiation sickness. A further 15 people perished from thyroid cancer caused by ingesting radioiodine. Despite aggressive monitoring, no increase was detected in solid tumors or mortality, even in the hundreds of thousands of minimally protected workers who purged the site after the accident. Nor did the data indicate any increase in birth deformities after the disaster. As the 2008 UNSCEAR report states: "There is no scientific evidence of increases in overall cancer incidence or mortality rates or in rates of nonmalignant disorders that could be related to radiation exposure."

How can we square this with our perception of the disaster's toll? The short answer is that we can't. In pushing their cause, many organizations were not above exaggeration or outright mendacity. The haunting photos of deformed babies beloved of Western charities were not products of Chernobyl but examples of normal deformities that occur at low levels in all populations, presented out of context. Greenpeace's huge death-toll estimate was conjured from bad science and wild extrapolation, prompting WHO spokesperson Gregory Härtl to note: "One always has to remind people why people make such estimates." Sacrificing facts on the altar of ideology doesn't merely muddy the waters, it's actively damaging to the psychological health of those afflicted. A 2006 WHO report warned that "Designation of the affected population as 'victims' rather than 'survivors' has led them to perceive themselves as helpless, weak, and lacking control over their future. This ... has led either to overcautious behavior and exaggerated health concerns, or to reckless conduct."*

Japan's Fukushima incident has become the focus of similar messaging. In March 2011, an earthquake off the Pacific coast of Tōhoku caused a deadly 50-foot-high (15 m) tsunami. The wall of water overwhelmed the Fukushima nuclear plant. With waterlogged diesel generators unable to cool the plant, a small leak of nuclear material ensued. The unfolding drama captured the world's attention, giving rise to a preponderance of breathless headlines. In reality, radiobiological consequences have been relatively minor; there has been only a single death linked to radiation exposure, and

---

* None of this detracts from the tragedy that 43 people succumbed to radiation needlessly. Some exposed in 1986 might yet exhibit some ill effect, though the passage of time has drastically diminished this likelihood. 115,000 were evacuated, and to this day an 18-mile (30 km) exclusion zone around the reactor has been maintained for precaution, despite the radiation level within this boundary being far below dangerous levels. Unmolested by human hands, the exclusion zone has transformed into a stunning wildlife habitat.

it is extraordinarily unlikely this figure will change drastically. The volume of radioactive material leaked is so minuscule as to be of little health concern. There is no detectable accident radiation in food grown locally, nor in fish caught nearby. These facts have been no impediment to the adoption of Fukushima as an anti-nuclear totem by campaigners, buttressed with dubious claims. Somewhere in the furor over Fukushima, we lost sight of the 16,000 lives lost in Japan to that cataclysmic tsunami.

Nuclear energy in isolation is, of course, not a panacea. It has intrinsic complexity and waste must be carefully contained. Nevertheless, by any objective metric, it is clean, safe, and hugely efficient. The opprobrium against it can be traced to the genesis of the modern environmental movement, where groups like Greenpeace began as a protest against nuclear-weapons testing. In the specter of the Cold War, an unfortunate conflation arose between nuclear weapons and nuclear power. Both became tarred with the same brush, despite operating on entirely different principles; one can no more turn a power plant into a thermonuclear bomb than one can turn a paper airplane into a fighter jet.* Persistent fearmongering over nuclear power has not only left an impression of exaggerated danger, it has blinded us to context. When the Banqiao hydroelectric dam failed in China in 1975, it killed over 171,000 people and displaced 11 million more. Even wind power has resulted in more than 100 deaths since the 1990s. This doesn't denigrate these vital technologies; the reality is that every form of energy production has risk.

Our reliance on fossil fuels is most costly to our environment and our health. Aside from being the high-carbon engine driving climate change, fossil fuels kill around 5.5 million people a year from air pollution alone. Following Fukushima, Germany acquiesced to demands from anti-nuclear campaigners to shut down its nuclear sector. Instead, they constructed heavily polluting fossil-fuel plants.

---

* The word nuclear itself has such negative connotations that it was dropped as a prefix to magnetic resonance imaging (MRI) lest it worried patients.

Japan also reduced its nuclear grid, in the process becoming the second-largest net importer of fossil fuels in the world. By 2017, Germany was the greatest carbon emitter in Europe, the phasing-out of nuclear power responsible for an extra 80 million tons of carbon dioxide in the atmosphere. France, by contrast, has long produced 78 percent of its energy through nuclear power, enjoying the cleanest air and among the lowest carbon emissions of any industrialized nation. The unintended consequence of dogmatic opposition to nuclear power is increased reliance on fossil fuels, accelerating climate change. If this is a "victory" for the environment, it is a deeply Pyrrhic one.

The crux of the problem is that unwillingness to yield to facts condemns us to terrible paths. Twisting reality to amplify one's convictions only serves to kill off any possibility of rational discussion, leaving us more divided and less informed. We cannot find pragmatic solutions to our problems if we refuse to be guided by the light of evidence. Ideology, like faith, has a nasty habit of recasting inflexibility as a virtue, dismissing anything not perfectly aligned with the tenets of that ideology. Voltaire's maxim that "the perfect is the enemy of the good" reflects the reality that there are often no ideal solutions; nor does sound reasoning always underpin ideological impasses. Inability to compromise or adapt often leads to poor outcomes. How can we confront existential challenges such as climate change if a substantial number of us deny its very existence while others undermine potential solutions? It's akin to living in a burning building where many of the residents refuse to accept it while others exert a dogmatic veto on calling the fire brigade.

If we are to survive and thrive, our opinions and beliefs must evolve with the facts. We can discuss and disagree on what the optimal solutions to our problems might be and how to achieve them, but we cannot get to that point if we insist on ignoring reality and substituting our own delusions instead. We are entitled to our own opinions, but not to our own facts. It would be bad enough were it only issues of science and health on which we are misled. But, as we have seen plenty of times already, dubious claims pollute political discourse, online and offline. Many of us dwell within the comfort of our echo chambers, seeking out sources that confirm rather than challenge our prejudices. As we become more starkly polarized than ever before, distinguishing fact from fiction is no easy undertaking. It's enough to drive anyone to apathy and cynicism. But apathy is the enemy; under its spell, we are dangerously pliable.

Nonetheless we are not doomed. Our greatest defense against all these challenges is the same virtue that has long driven our success as a species: our inquisitive minds. The ability to think analytically is a powerful shield against an onslaught of nonsense. Over the years, skeptics and scientists have done sterling work debunking purveyors of abject nonsense, from exploitative psychics to dangerous cranks. But perhaps the biggest challenge we face in the twenty-first century is the rise of conspiracy theories, infecting every arena from politics to medicine. Throughout this book, we've seen these ways of thinking manifest in myriad forms and the harm they can do. The narratives peddled by the prophets of paranoia are all-encompassing and seductive. In their wake, they sow discord and distrust, driving us further apart and rendering us vulnerable to harm. Like viruses, conspiratorial thinking evolves and mutates rapidly, acquiring strong immunity to reason along the way.

Yet they can only thrive by suppressing reason; unsurprisingly, conspiratorial ideation is associated with low levels of analytic thought. Evidence to date suggests that acceptance of such beliefs is strongly associated with a frugal, intuitive information processing—a tendency to go with rapid "instinct." Conversely, strong analytical thinking is associated with open-mindedness but negatively associated with belief in conspiracy theories, as this approach lends itself to critical evaluation of claims—especially those that are illogical or lacking in evidence. Consequently, those employing analytic thinking styles are much less likely to fall victim to the cognitive biases we've previously encountered. But crucially, this isn't set in stone—research indicates that eliciting analytical thought reduces conspiratorial ideation, even in groups prone to it. By engaging our capacity for analytic thought, we can liberate ourselves from the clutches of even the most noxious worldviews.

By simply being aware of potential errors in our reasoning, we can protect ourselves from detrimental consequences. Throughout this book, we've explored numerous trapdoors that we need to avoid, from logical to psychological to rhetorical and beyond. Being aware of these is half the battle—simply knowing about the pitfalls and traps we can fall victim to makes us less likely to blunder into them. The other half is applying this knowledge. To weigh an argument, we must not only consider how the reasoning flows but also interrogate the premises themselves. Are they well supported or do they pivot on rhetorical trickery? Are the conclusions drawn valid or dubious? This is precisely what we did with the cancer-conspiracy claim earlier on: by showing that the premises were flimsy, we opted to reject the claim. We saw too with fears about nuclear energy that popular perception on an issue is no substitute for critical thinking.

To answer the difficult questions we're faced with, we need to make use of the concept of scientific skepticism. At its core, this means asking the relevant questions to determine whether what we're presented with is reasonable or not. The word "skepticism" itself stems from the Greek *skeptomai*—to consider carefully. The philosopher Paul Kurtz defined a skeptic as "one who is willing to question any claim to truth, asking for clarity in definition, consistency in logic and adequacy of evidence. The use of skepticism is thus an essential part of objective scientific inquiry and the search for reliable knowledge."

Skepticism is implicit in the scientific method, the very lens we use to interrogate the universe. But it's every bit as fundamental to our political and societal health. Without it, we cannot hope to question the assertions of those in power or those seeking power. If we do not know to ask for evidence or what constitutes reliable information, we are powerless against the whims of the demagogues, dictators, and charlatans who would seek to exploit us. Without healthy skepticism, we are malleable to manipulation, weaponized to dire ends. Bereft of the protection against fanaticism that analytical thought brings, we are vulnerable to those who would deceive us. Our history is littered with reminders of just how terrible the consequences of this can be.

Skepticism implores us to seek truth rather than fool ourselves with comforting fictions. It demands that we follow evidence and logic through to their conclusions, whether we like those conclusions or not. This isn't always a comfortable experience; analytical thinking has led to the slaughter of many a sacred cow. But it is the only way to overcome our fallibility and blind spots.

So, when faced with a claim, how can we approach it in an analytical manner? We can't simply accept it because it confirms a preexisting prejudice; nor can we appeal to the wisdom of crowds to gauge it for us. Instead, we subject it to scrutiny to determine how much stock we ought to place in it. In the words of Carl Sagan, "extraordinary claims require extraordinary evidence." Over the course of this book, we've explored many aspects we need to consider when confronted with a claim, be it political, scientific, or otherwise. While it's rarely straightforward, a condensed checklist of things to ask might include:

- *Reasoning:* Do the premises lead to the conclusion presented or is something askew in the reasoning? To be valid, every link in the chain of argument must connect seamlessly to the others. A lurking non sequitur suggests something amiss. Similarly, if following the argument through to its logical conclusions yields contradictions or absurdity, it's a warning to be cautious. The premises themselves are vital too; are they reasonable and well supported or do they disintegrate under interrogation? If the premises wither in the light of inquiry, the conclusion that stems from them can usually be dismissed.

- *Rhetoric:* What kind of argument is being made? Authority alone is no substitute for evidence; if an authority is cited, it must provide evidence for the claims made. Narratives that reduce complex situations down to a simple cause ought to be considered with caution, as should those that force a complicated spectrum of views into an artificial binary. Attempts to misrepresent the position of others should be dismissed out of hand. The onus to prove a claim is always on the one asserting it, and approaches that rely solely on denigrating or smearing an opponent prove nothing.

- *Human factors:* What biases might be at play in different accounts? None of us is immune to instances of motivated reasoning or confirmation bias. Determining whether a position is reasoned or ideologically driven is imperative. Is the argument put forward based on cherry-picked information to support a particular point of view? We are in many respects the epitome of the unreliable narrator; when the evidence at hand is subjective or anecdotal, we cannot overlook the fact that perception and memory are imperfect.

- *Sources:* Where does the information come from? Does it come from reliable, verifiable sources? Assertions that cannot be traced back to a reliable source should not be seriously considered. The information we acquire is often shaped by our own echo chambers and ideology. We must take pains to verify whether it is fair-handed or merely reflective of what we want to hear. Anecdotal information must be very robustly assessed to gauge its merit. The mere fact that there are contested viewpoints does not mean every opinion is equally well supported.

- *Quantification:* Can the claim be quantified? If numbers are presented, the context for those figures is vital. Statistics are useful but they can be employed and manipulated to fool the unwary. The difference between relative and absolute risks must be kept in mind, and we must compare like with like.

And, as always, the mantra that correlation does not imply causation must never be forgotten.

- *Science:* Is the claim testable? Can it be falsified, at least in principle? If the claim presents a seemingly scientific hypothesis, is it based on reputable work or does it rely on cargo-cult science? If scientific data is presented, does it reflect the consensus view (totality of evidence) or cherry-picked outliers? Is the supporting data strong enough to support the conclusion? If the data can be equally well explained by another hypothesis with fewer assumptions, Occam's razor suggests caution.

This is a useful set of questions to ask when confronted with a new idea. The most important ideas we critique, however, are our own. To think like scientists, we must be willing to be guided by evidence and reason, to admit when we're wrong and rectify it. This means accepting that all conclusions and positions are provisional, and subject to change in the light of new information. This isn't easy for us—as we've seen, we are deeply attached to our beliefs, to the point where we often interpret a challenge to them as an attack on ourselves. But this is a flaw we must strive to confront.

Our ideas do not define us. They are often wrong, and there should be no shame in adapting to new information. To do so is laudable; the only shame is refusing to change our minds when evidence demands it. Nor should we feel any pressure to jump to an instant opinion when the evidence simply isn't there. Positions formed in haste are often wrong and resistant to change. There is no shame or cowardice in not leaping to a conclusion. Uncertainty might be uncomfortable, but we must endure it. As Bertrand Russell once warned, "So long as men are not trained to withhold judgment in the absence of evidence, they will be led astray by cocksure prophets, and it is likely that their leaders will be either ignorant fanatics or dishonest charlatans. To endure uncertainty is difficult, but so are most of the other virtues."

There is one final important thing to note: Skepticism should never be confused with cheap cynicism. It is not a knee-jerk "I doubt that" but rather "why do we think that?" It's an open process that encourages discussion and understanding, not a means to shut it down. By the same token, there are many contrarian writers and broadcasters who paint themselves as "skeptics" on issues such as climate change or vaccination. This is a calculated misnomer. Skepticism demands that all claims are treated as unproven until they're confirmed or falsified. Denialism, by contrast, is a stubborn and persistent refusal to accept what the evidence shows beyond all reasonable doubt. Self-proclaimed "skeptics" who oppose the overwhelming scientific consensus are really rank denialists, deliberately refusing to accept incontrovertible evidence that their position is untenable.

I would be remiss if I left you with the impression that the scientific method is merely a framework for debunking bogus claims, a perpetual rain on the parade. It is so much more; scientific inquiry is a burning torch that casts light on the encroaching darkness of ignorance and fear. The reality of the universe we inhabit is so much more astounding than any fiction we can concoct. Take, for example, the elements vital to life: carbon, oxygen, nitrogen. The knowledge that these elements can only be created inside massive stars leads inevitably to one conclusion: The atoms that constitute us were forged eons ago in the nuclear heart of an exploding star, expelled vast distances across the universe. We are quite literally stardust, born of the ashes of supermassive suns. Far from being stifling, instead reasoned thought yields discoveries we never dreamt possible. And it is all that stands between us and the perpetual darkness of ignorance.

EPILOGUE

Aime la vérité, mais pardonne à l'erreur. ("Love truth,
but pardon error.")

—VOLTAIRE

Choosing one's battles is vital, as the co-discover of evolution
Alfred Russel Wallace would attest. In January 1870, a gaunt-
let was thrown down in the journal *Scientific Opinion*, challenging
anyone to prove that the world was a sphere and not flat. The pro-
poser of this wager was John Hampden, a wealthy religious zealot
besotted with literal interpretation of the Bible. So assured was he
that scripture would affirm Earth was flat, he was willing to part
with a cool 500 pounds to anyone who could prove him wrong.
This seemed ridiculous—the spherical nature of the world had been
known since Greek antiquity. Eratosthenes had accurately calculated
the circumference of the planet three centuries before the Common
Era, and circumnavigation of the globe from the 1500s onward proved
it. Still, Wallace was in a precarious financial situation and intrigued
by the bet. He consulted with the geologist Sir Charles Lyell over
whether he ought to accept. "Certainly," Lyell replied. "It may stop
these foolish people to have it plainly shown."

Wallace and Lyell were of the mind that these men were simply
misguided, capable of being brought to reason by suitable demonstration. After some amicable correspondence, Wallace and Hampden

met at the Old Bedford Canal in Norfolk. The proposed experiment was straightforward: At two bridges six miles apart, Wallace would affix markers an equal height above water level. At the midpoint between the bridges, a pole would be erected with markers the same height above water. If Hampden was correct and the world was flat, the three markers would align when viewed through a telescope. If the world had a convex curve, then the midpoint marker would appear elevated when viewed through the scope. The experiment proved curvature, as might be expected. Unbeknownst to Wallace, though, Hampden was already attempting to load the dice in his favor, choosing a flat-Earth creationist as a referee. They objected on technicalities, so Wallace reran the experiment to their specification. Still Earth was curved, and Wallace was declared the winner.

This vindication proved a hollow victory, as Wallace's troubles were only beginning. Hampden refused to accept the result, mounting a protracted legal campaign to nullify the bet. While Wallace won several judgments, Hampden refused to pay, eventually declaring bankruptcy. Becoming increasingly unhinged, he began to write vitriolic and threatening letters; to Wallace's wife, he wrote: "Madam, if your infernal thief of a husband is brought home some day on a hurdle, with every bone in his head smashed to pulp, you will know the reason. Do you tell him from me he is a lying infernal thief, and as sure as his name is Wallace he never dies in his bed." The libel case aside, death threats were far beyond the pale for English law, and Hampden was eventually sent to prison. In Wallace's words, the affair

> cost me fifteen years of continued worry, litigation, and persecution, with the final loss of several hundred pounds. And it was all brought upon me by my own ignorance and my own fault—ignorance of the fact so well shown by the late Professor de Morgan—that "paradoxers," as he termed them, can never be convinced, and my fault in wishing to get money by any kind of wager. It constitutes, therefore, the most regrettable incident in my life.

Wallace's woeful experience predated the advent of satellite technology, space travel, and commercial aviation. One could be forgiven for assuming that such advances would have relegated flat-Earthers to the waste bin of history. But no—they have flourished online, insulated in communities that amplify their beliefs. In their forums, attempting to explain away the scientific evidence against them, they expound all manner of mechanisms that in general pivot on mangled understanding of both geometry and the phenomenon of refraction. And, as Wallace discovered, they are highly unlikely to change their views in the face of evidence.

They're not unique in this trait—shifting anti-vaccine views, for example, is remarkably difficult. A 2014 study in California found that refuting claims of a link between the MMR vaccine and autism paradoxically decreased intention to vaccinate among parents with the least favorable attitudes toward vaccination. For those resolutely opposed to vaccination, rational approaches further entrenched them in their ill-founded views.* It might be tempting to dismiss such views as fringe, but the reality is that we live in an interconnected world. It has never been easier to spread myths far and wide. Our concerns are especially potent in shaping how we understand the world around us. Misinformation and fearmongering can be marshaled by the most fringe of elements to wreak havoc on a much larger scale; and perhaps nowhere is this more heartbreakingly obvious than the worldwide confidence crisis over the human papilloma virus (HPV) vaccine.

The specter of HPV has long haunted humankind, exploiting perhaps our deepest drive: our insatiable libido. HPV transmits during sexual contact, and virtually all sexually active adults carry some of its more than 170 known strains. The majority are harmless or easily cleared by the immune system. But some variants

---

* Some would consider this an example of the "backfire" effect, where irrational beliefs become stronger in the face of contradictory data. The evidence base for the backfire effect is mixed, but we'd expect something similar to manifest at least some of the time due to motivated reasoning. On a tangent, Wallace brought himself into disrepute by being actively anti-vaccine—further evidence that brilliance in one area alone does not overcome one's ideological blind spots.

are more ominous; subtypes 16 and 18 can lead to a family of cancers.* HPV infection is responsible for approximately 5 percent of cancers worldwide, including over 90 percent of cervical cancers, which alone claims approximately 270,000 lives each year. The HPV vaccine was a revelation, capable of banishing this vicious ghost forever to the dim haze of memory. Gardasil provided protection against the most odious subtypes, and by 2007 was licensed in more than 80 countries. Results were astounding: by 2013, the vaccine had resulted in a staggering fall of 88 percent in HPV infection for American girls aged 14 to 19. By 2018, Australia stood on the cusp of eradicating HPV in young women due to vaccination. For the first time in history, we could eradicate an entire family of cancers.

But it is impossible to immunize against foolishness. In America, religious conservatives thwarted a national vaccination drive, concerned that the vaccine encouraged wanton promiscuity. Their presumption that vaccination is a passport to unadulterated sexual abandon, however, doesn't withstand even basic scrutiny; evidence shows sexual activity is not elevated in vaccinated cohorts. The irony of preaching abstinence in lieu of protection is that it simply doesn't work—teens subjected to this approach begin sexual activity at the same stage as other teens. Whether intended or not, a sizeable contingent of the American public found the idea of their teenagers having normal sexual urges so profoundly disquieting that they were willing to risk their children's needless deaths rather than act pragmatically.

But the age-old problem of anti-vaccine activism was to prove a far greater problem. Immediately, anti-vaccine campaigners attributed a kaleidoscope of adverse reactions to Gardasil, including a constellation of nebulous, inconsistent, and subjective symptoms. These assertions were, however, completely unsupported by epidemiological data; extensive adoption of the vaccine around the world had yielded follow-up studies involving millions

---

* Subtypes 6 and 11 also cause genital warts, which the vaccine protects against too.

of women. In these vast data sets, even rare side effects would be exposed. Yet all evidence showed the vaccine was safe, well tolerated, and immensely effective at preventing HPV infection. But the anti-vaccine movement has never been concerned with reality, only ideology. And with a new cause célèbre, this fringe but vocal band targeted their misinformation at politicians and parents on social media worldwide.

The consequences were severe. By 2013, a panic had erupted in Japan, culminating in the health minister suspending the vaccine's recommendation. Subsequent investigation swiftly concluded that the vaccine had nothing to do with reported ailments, but it remained political poison; the vaccination rate crashed from 70 percent to less than 1 percent by 2017. By 2014, vaccine-damage tales materialized in Danish media, following intense lobbying by anti-vaccine activists. Their impact was exacerbated by an inflammatory TV2 feature implying the vaccine damaged young women, with emotive testimonials mere facts and figures couldn't overcome. Self-diagnosis of "vaccine damage" became common and unquestioned, and vaccinations dropped precipitously, falling from 79 percent to just 17 percent.

In 2015, the panic arrived in Ireland. At the time I was based at the University of Oxford but closely followed affairs in my home country. Vaccination and cancer were topics I frequently covered, so it wasn't unusual that my opinion was sought. But suddenly I was inundated by an influx of journalists with questions about the ostensible dangers of HPV vaccination. They had been told alarming things: that the vaccine contained toxins, hadn't been tested properly, and that ill effects were being covered up in a medical–industrial conspiracy. Perhaps the most important lesson one learns at the coalface of science communication is that not all ostensible concerns are in good faith. It is completely understandable that people have questions about vaccine safety; it is another thing entirely to engineer or exploit public uncertainty in order to fearmonger. While the situation in Japan and Denmark hadn't yet filtered into anglophone media, the claims put before me bore the unmistakable hallmarks of anti-vaccine propaganda.[*]

Not only were these assertions profoundly wrong, they weren't even new. They were staples of the anti-vaccine canon given a cursory dusting-off, zombie myths that persisted despite thorough debunking. They retained currency in only one domain: the underbelly of the anti-vaccine community. Their repetition now was an alarm bell warning that this ostensible concern was tainted by the fingerprints of anti-vaccine activism.

I pointed to the vaccine's tremendously positive safety profile, emphasizing its ability to obliterate an entire family of cancers. This dissuaded most of the reporters I spoke with from giving the story oxygen; as I've alluded to before, sometimes the best contribution a scientist can make to public understanding is to kill dubious stories before needless panic takes root. But less diligent outlets had already run the story of a group whose name kept popping up: REGRET. Their acronym laid their beliefs bare: "Reactions and Effects of Gardasil Resulting in Extreme Trauma." They claimed to represent hundreds of young women damaged by the vaccine, many allegedly so ill they were confined to a wheelchair or under 24-hour suicide watch due to the agony they endured. The group was omnipresent on social media, targeting their message at politicians across the political spectrum.

They also had a penchant for headline-grabbing stunts. August 2015 saw them stage a protest at an Irish Cancer Society (ICS) talk in Galway, culminating in the verbal abuse of world-renowned virologist Professor Margaret Stanley. Such was the intensity of the ordeal that Stanley commented she had "never experienced in my professional life the vitriol and animosity that was expressed."

---

[*] Aluminum toxicity, for example, was repeatedly mentioned. This thoroughly debunked trope came to prominence during the Wakefield debacle and asserts that the minuscule concentrations of aluminium used in some vaccines cause autism. Leaving aside the fact that autism isn't an acquired condition one can "catch," such logic would suggest people would become neuro-atypical every time they cut themselves opening a can. Adding insult to injury, aluminum wasn't even used in the MMR vaccine.

A deluge of threats forced the ICS to arrange security for future events. Coordinated targeting of critics became standard fare. After debunking myths on radio and in the press, broadcaster Ciara Kelly and I were inundated with vexatious complaints. This wasn't the first time I'd endured such tactics, and I was lucky to have a supportive university. Ciara was a practicing GP, however, and her detractors lobbed multiple accusations against her to the Irish Medical Council, each obliging a stressful investigation. As Ciara noted, she had "never had a complaint from an actual patient–all of my dealings with the medical council came from anti-vaccine complainants I've never met. While extremely stressful, I was absolutely determined that an anti-science agenda could not be given in to."

Intimidation tactics aside, the narrative of dedicated parents seeking answers held a powerful allure. Initial stories were sympathetic, bereft of journalistic skepticism, despite even a cursory glance at REGRET's online presence revealing an anti-vaccine theme far beyond Gardasil.* This didn't evade everyone's attention; journalist Susan Mitchell asked whether she might speak with medical professionals to verify the claims made, questioning how the considerable sums raised from public donations were being managed. Faced with a diligent journalist unwilling to write another puff piece, REGRET declined to respond.

A fearmongering documentary by broadcaster TV3 in late 2015 had an immediate and drastic impact on public opinion; use of the vaccine, which had stood at 86.9 percent in 2014, plummeted to about 50 percent by 2016. A procession of politicians lined up to cast doubt on the safety of the vaccination; one blowhard Irish senator insisted that "14-year-old girls don't lie en masse." That every deleterious effect attributed to the vaccine was more readily explained by common psychological† and physical illnesses was thoroughly

---

* One founding member proudly boasted of having never immunized her five children, directing "vaccine-injured" girls to her husband's homeopathic practice for treatment.

† Much like the panic over "Wi-Fi damage," the nocebo effect seems likely to have played a part too–particularly with impressionable teenagers.

ignored. In the court of public opinion, the vaccine was no longer seen as vital but dangerous.

In response, the National Immunization Office quickly established a steering group of concerned organizations, from medical associations to parental groups. The newly forged coalition produced clear information packages for parents and health professionals, disseminating them widely across social media to provide an authoritative counter to the dominant falsehoods. The health minister and senior politicians lent unconditional support, reiterating the vaccine's safety and necessity—a united front utterly lacking in Japan and Denmark. But information could only be part of the battle; the emotional front was the other.

Fear and uncertainty are powerful motivators. The vast majority of those not vaccinating their children weren't dyed-in-the-wool anti-vaccine zealots but simply parents eager to do what was best for their children. A small core of anti-vaccine activists had dominated the narrative, perpetuating a received wisdom that the vaccine was unsafe. The ringleaders had effectively exploited the empathy of those unfamiliar with their tactics. Under the guise of seeking answers, anti-vaccine activists garnered sympathetic coverage, downplaying the fact that repeated investigation worldwide did not support their claims. The catch-22 was that anti-vaccine activists weaponized this as "dismissal," denigrating scientific and medical professionals as an unfeeling morass of vested interests. Unsure of what to believe and unclear about the benefits, parents opted to stay away. In the sound and fury, the rationale behind the vaccination was erased from conversation. But the real choice wasn't between phantom side effects or none; it had always been between protecting one's children from an awful family of cancers or risking their lives over a fiction.

Still, the depth of feeling on the issue was such that merely defending the vaccine in the public sphere attracted furious commentary across social media—storms of personal abuse, allegations of corruption, and, most often, the charge that we simply "didn't care" about people. The cruel irony is that this aspersion couldn't have been more wrong. Beneath every health drive lies a genuine desire to save lives, motivated by awareness of a sad reality; behind every mortality statistic is a tragedy, a family torn asunder, a loved one lost. Far from indifference, the vociferous advocacy by the health care and scientific communities reflected the fact that the vaccine spared so much preventable misery. Empathy is universal—no decent human being requires an ulterior motive to save lives or reduce the suffering of others. For vaccine use to have any hope of recovering, reframing the narrative was imperative.

Research by the Health Service Executive (HSE) suggested individual stories held disproportionate sway in shaping intention to vaccinate—precisely why the emotive yet unsubstantiated personal accounts from anti-vaccine groups garnered more traction than facts alone. In August 2017, the HSE launched a campaign featuring young vaccinated women, urging audiences to "protect our future." Toward the year's end, world experts on HPV elimination convened in Dublin under tight security. I was there to speak about countering misinformation, mindful that other countries afflicted by a confidence crisis had seen no hint of recovery. It remained to be seen whether Ireland's approach would yield dividends. Mid-meeting, the data arrived: Vaccinations had climbed to 62 percent—an encouraging suggestion that well-aimed messages could overcome the fog of misinformation.

Negating the toxic influence of fearmongering on public understanding meant control of the narrative had to be wrested from those who thrived on fear. The availability heuristic, though, exerted powerful influence. While stories of vaccine-damaged young women lacked substance, persistent fearmongering had branded them onto the public consciousness. Shifting perception required an extraordinarily powerful account to reframe the narrative, laying bare why vaccination mattered. The HSE didn't have to go looking for this; it came to them. Laura Brennan was only 25 when she was diagnosed with metastatic cervical cancer. Facing a difficult prognosis, she offered to share her experience to help others. It would be difficult to envision a better advocate. Articulate, charismatic, and beautiful, Laura was an antidote to the poisonous fictions swirling around the vaccine, and the campaign placed her words to the fore:

> At 24, I was diagnosed with cervical cancer stage 2B. I was quite optimistic, as there was something that could be done. With chemo and radiation, there was a good chance it could be cured. Two months later it was back—and things are different this time. There is no treatment that will cure my cancer; there is only treatment that will prolong my life. If anything good comes out of this, I would hope parents get their daughters vaccinated. The vaccine saves lives—it could have saved mine.

It is impossible to be unmoved by Laura's bravery and frankness. She reframed the vaccine not as a danger, but as a bulwark against tragedy. Not only did this provide a counter to the emotional narratives expounded by anti-vaccine campaigners, it also had something they lacked: scientific facts. The campaign melded Laura's story with expert opinion, giving a consistent message hugely effective at rebutting the fearmongering that had until then dominated. Prior to this, there had been no obvious human element to combat the emotive fearmongering, but Laura's bravery and strength was the linchpin of the fight against misinformation, presenting the importance of the vaccine in stark terms. So effective was this approach that international health bodies ultimately adopted it, and Laura went on to front the WHO's worldwide HPV vaccine campaign too.

While anti-vaccine elements didn't suddenly dissipate into the ether, this shattered their stranglehold on the conversation. In July 2018, vaccine coinventor Ian Frazer was keynote speaker at a meeting where Laura and I were also speaking. REGRET protested outside, but public sympathy toward them had largely evaporated. For the first time since the panic broke, journalists present were far more interested in the content of the meeting than in the antics outside—Laura's story especially. Upon leaving, we were heckled by protesters decrying the vaccine. Unfazed by the intimidating atmosphere, Laura's rebuke was unanswerable: "If I'd had the vaccine, I wouldn't have cancer." As her cancer was caused by HPV-16 infection which the vaccine protects against, this was undoubtedly true.

Changing minds is vital, but hearts matter every bit as much; we are not intellectual automatons, but emotional creatures who feel first and think later. All the facts, arguments, and logic in the world are for nothing if we cannot connect on an emotional level. Laura's story saved more lives than a library of journal articles in isolation ever could. Her contributions to rising vaccine use have been recognized by everyone from the Royal College of Physicians in Ireland to the WHO, who adopted her story for their vaccine campaign. This is even more admirable considering the great personal strength required, and the needless stigma and sexual connotations existing around HPV. As Laura told me:

> It is incredible that we have a vaccine that protects against cancer. Misinformation frustrates me, as it is a scientific fact the vaccine is safe and saves lives. If sharing my story changes one parent's mind about getting their child vaccinated, it's a potential life spared from going through what I've gone through. I will continue to use my voice at every opportunity I'm given, in the hopes that the next generation won't have to suffer like me and so many others have. I'm a terminally ill woman dying of a cancer that is now preventable; why wouldn't you want to protect your child from this is the question every parent should ask themselves.

I was privileged and honored to count Laura as a close friend. She slipped away on March 20, 2019, aged 26. Her death elicited a huge outpouring of public grief—and laid bare the tragedy the vaccine could prevent. The anti-HPV vaccine debacle encompasses many familiar themes: the triumph of anecdote over data; the media's influence on perception; the impact of motivated reasoning. But it also reminds us that, while evidence and reasoning are vital, so too is emotional framing. To change minds and hearts, we must not only offer better arguments, but remind people on a visceral level why it matters. Laura's powerful advocacy refocused the world's attention on precisely why the vaccine was so vital. Vaccinations increased over 20 percent during her 18 months of campaigning, and by selflessly working with the WHO and HSE, her impact was felt worldwide—an enduring legacy for a brilliant woman. Just mere months after Laura's untimely death, the vaccination rate in Ireland climbed above 80 percent—and in Laura's home county of Clare, it has exceeded 90 percent. Even in death, Laura will continue to save untold lives.*

We cannot forget that we are social animals, disproportionately influenced by the opinions and positions of those around us. We coalesce into tribes bounded by a shared worldview—even when that worldview is supremely misguided. Conspiracy theorists, for example, tend to operate in polarized echo chambers, closing them off to other sources of information as they become more invested in the narrative. These aren't just beliefs; they're part of something more—an identity, with the allure of special knowledge and control—even if it is illusory. To believers, this not only simplifies the complexity of the world, it also provides an ego boost. Research indicates that believers consider themselves part of a special in-group, superior to the deluded masses, eager to stand out from the crowd of "sheeple." Incidentally, the curious rise of the anti-mask movement in the midst of a pandemic is a stellar example of precisely this. Despite the infectiousness of COVID-19, anti-mask protests draw considerable crowds, such as the 17,000 people who gathered in Berlin in August 2020, jeopardizing not only their own health but that of others. Similar scenes have played

---

* At the time of writing this is incredibly recent and raw. The hearts of those who loved her are truly shattered—mine included. But I draw some consolation from the fact that thanks to her selflessness, others will not have to endure the ordeals she went through, nor the loss we feel now.

out the world over; a study of French anti-maskers found that while they were drawn from across the political spectrum, they were united in a conception of themselves as "free thinkers." One might, of course, question the utility of face masks in different circumstances, but the opprobrium displayed towards them by opponents transcends our pretense of good-faith concern; urban legends that they deprive wearers of oxygen, for example, are easily debunked by a simple experiment with a mask and a pulse oximeter. Even by simply observing that armies of health-care workers have soldiered through the pandemic masked to the hilt for hours on end, we see none of them dropping dead due to oxygen depletion. Apparently, this superior self-perception doesn't map onto the most basic powers of observation. It doesn't matter how educated or experienced others might be—ultracrepidarians* are always convinced of their own rectitude, speaking with a confidence inversely proportional to their understanding.

Surrounding oneself with those who echo a belief reinforces it, insulating it from criticism. In this crucible, belief becomes the unifying element. To question elements of doctrine is to risk being ostracized from your tribe; those brave enough to renounce conspiracy theories, for example, are often set upon by their one-time associates for their apostasy. The grim reality is that there are always cohorts so ideologically invested in a belief that their convictions are impervious to the intrusions of reality. This might make things seem hopeless; if there are people for whom evidence means nothing, how can we ever come to consensus on difficult issues?

This sense of despair is magnified each time one ventures into the maelstrom of comment sections, but there's a ring of truth to the adage about empty vessels making the most noise: an analysis of comments on *The Guardian* website found that, at most, 0.7 percent of readers left comments, with 17 percent of all comments

* An ultracrepidarian is defined by the *Oxford English Dictionary* as one who expresses "opinions on matters outside the scope of one's knowledge or expertise." If there is a formal process for adopting collective nouns, I humbly suggest that "a twitter" be adopted for "ultracrepidarians."

attributed to 0.0037 percent of the readership. Other analysis indicates that many of those people commenting on articles don't bother reading them first. Vocal but minuscule cohorts are unlikely to be representative of opinion at large; the hellish state of online discourse gives an impression that everyone is locked in some binary state of war, but this describes only the most extreme. That the loudest voices are often the least informed and most partisan isn't a new problem; the poet W. B. Yeats lamented that "the best lack all conviction, while the worst / Are full of passionate intensity."

Most of us do not hold such entrenched or polarized positions. Of course, there will always be those so invested in their religious or political ideology that they would sooner ignore reality than adjust their views, or propagate myth rather than admit error. There are none so blind as those who will not see; arguing with the committed is an exercise in futility. But the vast majority are not beyond reach, amenable to both reason and nuance. Affecting change doesn't require swaying the entire world, only shifting the conversation toward evidence and reason. What matters is that we can distinguish between reliable information and that of which we should be wary. The comments that follow my own articles often descend into gladiatorial combat between those already in agreement with the article and those resolutely unwilling to consider the evidence presented. But these extreme ends of the distribution are not the intended audience; my efforts are focused on the silent majority who wish to understand, who seek something reputable amid a storm of dubious sources.

We are often told we live in a post-truth society, where outrageous falsehoods eclipse factual accounts. This is understandable; as I write this, the world is still reeling from the aftermath of Trump's victory and the UK vote to leave the European Union. Both campaigns were epitomized by outright mendacity, propaganda, and inflammatory fictions. But, although such events have blindsided us, people have not abandoned their fundamental desire for truth. We still have our capacity for curiosity and our urge for understanding. The real challenge in this era of instantaneous information is to distinguish between the reputable and dubious, to reflect rather than react, and to question vigorously what we're told. This has never been more urgent nor more difficult. The satirist Jonathan Swift remarked: "Falsehood flies, and truth comes limping after it, so that when men come to be undeceived, it is too late; the jest is over, and the tale hath had its effect." When Swift wrote this in 1710, it might have been hyperbole, but three centuries later it seems prescient. We are surrounded by an army of propagandists, charlatans, and fools who are determined to project fictions further and faster than ever before. Unchallenged, their machinations leave us pliable, prone to poor decisions and disaster.

The first step to combatting this starts with us. Our sense of identity is so entangled with our values and convictions that we can overlook something vital: We are not our ideas. We are not defined by beliefs, but by our ability to think. To be human is to err, blessed with the capacity to correct ourselves. There is no shame in being wrong, only in refusing to rectify our mistakes. We must be willing to adapt in the face of new information, to jettison wrong-headed beliefs when required, and to embrace truth even when unpleasant. Indeed, it is perhaps even more imperative that we question ourselves as rigorously as we would question others. The spirit of how we question matters too; we need to guard against motivated reasoning, and subject even our comforting beliefs to the same scrutiny to which we would hold opposing views.

The unspoken truth is that no one changes anyone else's mind; we can only change our own, giving others the tools and freedom to do the same. Moving toward an evidence-based society is a marathon, not a sprint. No one event upends deeply held misconceptions or ushers in automatic enlightenment. It must be a gradual process, where we absorb new information, correct our mistakes, and move toward a more informed view of things. We need to realize that it is not about knowing the "correct" answer, but of following the

correct method to arrive at a better understanding. In the pursuit
of making better decisions, it is preferable to be wrong for the
right reasons than right for the wrong ones, because ultimately to
embrace reason ensures we more often choose better paths than
blind luck or happy accident alone would ever allow. To bring forth
a more informed, healthier, and more equitable world, discussion
is imperative to stem the creep of falsehood, pseudoscience, and
dubious reasoning that tear us asunder.

We have long implicitly accepted debate as the arbiter of truth;
however, debate often rewards not the best arguments but the most
devious orators. In adversarial settings, it is rhetorical dexterity or
the ability to inflame an audience that frequently triumphs over
clarity and reason. The process itself often veers into false dichotomy, whittling spectrums of opinion down to two caricatured
counterpoints, forcing us to choose a side when the reality might
be somewhat more complicated. The process is intrinsically polarizing, rendering the act of changing one's mind or compromising
upon reflection impossible. Too often, debate makes us more
divided and less informed.

It is also the unwitting ally of false equivalence; I have lost count
of the number of times I have been invited on shows to "debate"
climate change or vaccines. As I tell producers, this is nonsensical
given that these are factual matters, no more suitable for "debate"
than the existence of Greenland. Political extremists and bigots
abuse the platform it provides, knowing full well that mere
exposure of their odious ideas is itself a victory. Pseudoscientists
and fringe groups crave debate, for it lends a veneer of legitimacy
to untenable beliefs and claims. This isn't to say that such topics
shouldn't be discussed—quite the contrary, they absolutely should.
It is vital we understand why climate change is real, tackle vaccine fears, or explore the resurgence in political extremism. But
discussion here is the operative term—the superficial, pugilistic
nature of debate forces us into rigid positions not conducive to
understanding. Discussion, by contrast, is a fluid process, where

our views can and should evolve. This latter dialectical approach encourages us to engage on an intellectual and human level, asking "why do you think that?"—and often more importantly, "why do I think that?" Conversation rather than combat changes minds and corrects errors—our own as much as those of others. In the words of Voltaire, we must "love truth, but pardon error."

Compassion is also incredibly important. We tell stories to understand the world, simple narratives where heroes and villains are clearly defined. We rush to deify or vilify, to sort people and events neatly into "good" or "bad." Life is rarely so clear-cut. We are all flawed and complex, capable of harboring foolish notions. These might be irrational, hurtful, or even hateful. Such is our fixation with essentialism that we attach labels to people rather than ideas. A point-scoring mentality that underpins modern discourse tends to descend into outright misrepresentation of differing positions, and often complete dehumanization of opponents.

But defeating straw men is meaningless. If we truly care about informed discussion, we should employ the principle of charity, which insists that we interpret an opponent's argument in the strongest, most rational way possible. Doing so pushes us to actively consider the perspective of others, giving us the means to either construct a thoughtful and robust counter or compelling us to modify our own views upon reflection. This doesn't mean excusing bigotry or rationalizing the indefensible, but simply ensuring we've been rigorous in our own thinking.

Ultimately, denigrating others is counterproductive; very few people change their mind when abused or dehumanized. I've found myself in the past using a contemptuous, haughty tone that I retrospectively loathe. I strive to avoid it because, while it may garner accolades from the like-minded, it risks alienating those who could benefit the most from any insight I might offer. Things are also usually more nuanced than most narratives allow. The pendulum of public opinion oscillates wildly, and those who would place you on a pedestal will as quickly cast you into the flames. But all of us at some point are misguided about some things. If we truly want a better world, we must allow others the freedom to evolve their views, without casting aspersions on their fundamental humanity— just as we ourselves should be afforded this compassion.

There are, however, caveats to such magnanimity. First, it only applies to good-faith discussion; those deliberately misrepresenting reality are unlikely to change their mind, and engagement with them is not likely to be constructive. Second, the ideal of open discussion should never be a cover for hatred or oppression. There is no onus on us to engage with hateful philosophies, and nor are we obliged to give a free airing to positions that deny others their fundamental rights or basic humanity. The paradox of tolerance is that a society tolerant without limit will eventually be overtaken by the intolerant. Karl Popper suggested that "we should therefore claim, in the name of tolerance, the right not to tolerate the intolerant."

Society itself is a fragile fabric, easily torn apart by misconception or fearmongering. We share a glorious world, and our fates are intertwined with bonds that cannot be severed. If this world burns, so do we all. We cannot hope to improve things if we labor under delusion and unthinking tribalism. Those who would subvert our thinking can make us deny reality, creating a vacuum that tyrants and charlatans fill with hatred and falsehood. Voltaire's warning that those who can make us believe absurdities can drive us to atrocity remains true, but the corollary to this dictum is equally important: Those who can erode human trust and cast doubt on shared truths can make us malleable to all evils. Whether this is propaganda that aims to sow discord, or misinformation propagated by those ideologically blind to reality, the net effect is societal division and distrust. Divided, we are weak and ineffectual, drifting toward disaster, incapable of collaborating on the truly global problems we face.

To allow facts, evidence, and reason to be disavowed is to stand on the precipice of tragedy. Berlin is home to many harrowing memorials marking the barbarity of the Nazi era. To me, the most unsettling one is the most understated. In beautiful Bebelplatz, there is a transparent floor plate in the center of the square. It commemorates the first Nazi book burnings, on May 10, 1933, where works deemed contrary to Nazi teachings were put to the flame. Today, this plate serves as a reminder of that madness; to glance downward is to be greeted by the haunting sight of row upon row of barren shelves, devoid of a single book. Inscribed close by are the words of poet Heinrich Heine: *Das war ein Vorspiel nur, dort wo man Bücher verbrennt, verbrennt man am Ende auch Menschen* ("That was only a prelude; where they burn books, they will ultimately burn people").

That monument in Berlin is a potent reminder of what dark consequences can arise when truth is sidelined and destroyed. Heine's words were written more than a century before Hitler seized power. He couldn't have envisaged the brutality of the Third Reich, nor how percipient his sentiment would prove. But he alluded to a fundamental darkness in those who would seek to erase truth rather than embrace it. There will always be those who would render us pliant with confusion and lies, but we are more resilient than we know. Even in this era where falsehoods perpetuate faster and further than ever before, our capacity for analytical thought is the blade that cleaves the reliable from the ridiculous. This can seem overwhelming, making a retreat into apathy tempting. But apathy is the enemy; we cannot challenge falsehood if we are disengaged, nor strive toward a better world if stricken by inertia. Only our willingness to question—to ask "why?" and "why not?"—shields against those who would mislead or manipulate us, a compass to steer us toward viable solutions to the challenges we face together.

These challenges are truly daunting, from climate change to antibiotic resistance to global pandemics and geopolitical instability. To meet them and endure, we need to think like scientists, reflecting before we react, guided by evidence over emotion, and always self-correcting. Striving toward a better future for all of us requires bravery and compassion as much as intellect. For, although we might start as mere irrational apes, we are endowed with the ability to be so much more. We must be unafraid to let go of poor ideas or embrace new ones. We must be forgiving not only of the errors of others, but also of our own. Ultimately, whether we prosper or perish comes down to whether we choose to learn from our mistakes or succumb to them.

.

REFERENCES AND FURTHER READING

Introduction

Rosling, Hans. *Factfulness*. New York: Flatiron Books, 2018.

Eco, Umberto. "Ur-Fascism." *New York Review of Books*, June 22, 1995. nybooks.com/articles/1995/06/22/ur-fascism/.

Wineburg, Sam, et al. "Evaluating Information: The Cornerstone of Civic Online Reasoning." Stanford Digital Repository, 2016. purl.stanford.edu/fv751yt5934.

Gabielkov, Maksym, et al. "Social Clicks: What and Who Gets Read on Twitter?" *ACM SIGMETRICS Performance Evaluation Review* 44(1) (2016): 179–92. doi:10.1145/2964791.2901462.

Hofmann, Wilhelm, et al. "Morality in everyday life." *Science* 345 (6202) (2014): 1340–43. doi:10.1126/science.1251560.

Brady, William J., et al. "Emotion shapes the diffusion of moralized content in social networks." *Proceedings of the National Academy of Sciences* 114(28) (2017): 7313–18. doi:10.1073/pnas.1618923114.

Vosoughi, Soroush, et al. "The spread of true and false news online." *Science* 359(6380) (2018): 1146–51.

Office of the Director of National Intelligence. "Assessing Russian activities and intentions in recent US elections," Unclassified Version (2017). dni.gov/files/documents/ICA_2017.01.pdf.

Paul, Christopher, and Miriam Matthews. "The Russian 'Firehose of Falsehood' Propaganda Model." RAND (2016): 2–7. rand.org/pubs/perspectives/PE198.html.

Hasher, Lynn, et al. "Frequency and the conference of referential validity." *Journal of Verbal Learning & Verbal Behavior* 16(1) (1977): 107–12. doi:10.1016/S0022-5371(77)80012-1.

Goertzel, Ted. "Belief in conspiracy theories." *Political Psychology* 15(4) (1994): 731–42. doi:10.2307/3791630.

Stanovich, Keith E. "Dysrationalia: A New Specific Learning Disability." *Journal of Learning Disabilities* 26(8) (1993): 501–15. doi:10.1177/002221949302600803.

Morewedge, Carey K., et al. "Debiasing Decisions: Improved Decision Making with a Single Training Intervention." *Policy Insights from the Behavioral and Brain Sciences* 2(1) (2015): 129–40. doi:10.1177/2372732215600886.

1. An Indecent Proposition

Nisbett, Richard, and Lee Ross. *Human Inference: Strategies and Shortcomings of Social Judgment*. Eaglewood Cliffs, NJ: Prentice-Hall (1985).

Federal Emergency Management Agency. "World Trade Center Building Performance Study: Data Collection, Preliminary Observations, and Recommendations. FEMA 403 (May 2002). fema.gov/pdf/library/fema403_cover-toc.pdf.

National Institute of Standards and Technology, U.S. Department of Commerce. *Final Reports from the NIST World Trade Center Disaster Investigation* (2011).nist.gov/el/final-reports-nist-world-trade-center-disaster-investigation.

Dunbar, David, and Brad Reagan, eds. *Debunking 9/11 Myths: Why Conspiracy Theories Can't Stand Up to the Facts*. New York: Hearst Books (2006; updated ed. 2011). skepticalinquirer.org/2011/the-conspiracy-meme/.

Goertzel, Ted. "The Conspiracy Meme: Why Conspiracy Theories Appeal and Persist." *Skeptical Inquirer* 35(1) (January/February 2011).

Grimes, David Robert. "On the Viability of Conspiratorial Beliefs." *PloS One* 11(1) (January 26, 2016). doi:10.1371/journal.pone.0147905.

Grimes, David Robert, et al. "Establishing a taxonomy of potential hazards associated with communicating medical science in the age of disinformation." *British Medical Journal Open* 10 (2020). orcid.org/0000-0003-3140-3278

2. Stripped to the Absurd

Hardy, G. H. *A Mathematician's Apology*. Cambridge, UK: Cambridge University Press (1992). Reprinted 2018 by Martino Fine Books, Eastford, CT.

Singh, Simon. *The Code Book: The Evolution of Secrecy from Mary, Queen of Scots to Quantum Cryptography*. Garden City, NY: Doubleday (1999).

Russell, Bertrand. *A History of Western Philosophy*. Collector's Edition. London: Routledge (2013).

World Health Organization, "Electromagnetic fields and public health: mobile phones." (October 8, 2014). who.int/news-room/fact-sheets/detail/electromagnetic-fields-and-public-health-mobile-phones.

INTERPHONE Study Group. "Brain tumor risk in relation to mobile telephone use: results of the INTERPHONE international case-control study." *International Journal of Epidemiology* 39(3) (June 2010): 675-94. doi:10.1093/ije/dyq079.

Frei, Patrizia, et al. "Use of mobile phones and risk of brain tumors: update of Danish cohort study." *British Medical Journal* 343 (October 2011): d6387. doi:10.1136/bmj.d6387.

Schüz, Joachim, et al. "Cellular phones, cordless phones, and the risks of glioma and meningioma (Interphone Study Group, Germany)." *American Journal of Epidemiology* 163(6) (March 15, 2006): 512-20. doi:10.1093/ajekwj068.

World Health Organization, op. cit.

Scientific Committee on Emerging and Newly Identified Health Risks. "Potential health effects of exposure to electromagnetic fields (EMF). European Commission (2015). ec.europa.eu/health/scientific_committees/emerging/docs/scenhir_o_041.pdf.

Danker-Hopfe, H., et al. Recent Research on Health Risk, Twelfth Report from Radiation Safety Authority's Swedish Scientific Council on Electromagnetic Fields, 12th report from Swedish Scientific Council on Electromagnetic Fields, 2017. edoc.unibas.ch/64887/.

U.S. Food & Drug Administration. Review of Published Literature between 2008 and 2018 of Relevance to Radiofrequency Radiation and Cancer (February 2020). fda.gov/media/135043/download.

International Commission on Non-Ionizing Radiation Protection. Guidelines for limiting exposure to electromagnetic fields (100 kHz to 300 GHz). *Health Physics* 118(5) (2020): 483-524. icnirp.org/en/publications/article/rf-guidelines-2020.html.

Grimes, David Robert, and Dorothy V. M. Bishop. "Distinguishing polemic from commentary in science: Some guidelines illustrated with the case of Sage and Burgio (2017)." *Child Development* 89(1) (January 2018): 141-47. pubmed.ncbi.nlm.nih.gov/29266222/.

Broad, William J. "Your 5G Phone Won't Hurt You. But Russia Wants You to Think Otherwise." *The New York Times* (May 12, 2019). nytimes.com/2019/05/12/science/5g-phone-safety-health-russia.html.

Ronson, Jon. *So You've Been Publicly Shamed*. New York: Riverhead Books (2016).

3. It Does Not Follow

Greene, Jeremy A. "For Me There Is No Substitute': Authenticity, Uniqueness, and the Lessons of Lipitor." *AMA Journal of Ethics*, 12(10) (October 2010): 818-23. doi: 10.1001/virtualmentor.2010.12.10.msoc2-2010.

United States Bureau of Chemistry, Service and Regulatory Announcements, Issues 21-30 (1917).

Wiseman, Richard, and Donald West. "An experimental test of psychic detection." *Police Journal* 70(1) (1997): 19-25. doi:10.1177/0032258X9707000104.

Druckman, Daniel, and John A. Swets. *Enhancing Human Performance: Issues, Theories, and Techniques*. Washington, DC: National Academies Press, 1988.

4. The Devil in the Details

Bing, Franklin C., "Vitamin C and the Common Cold." *Journal of the American Medical Association* 215(9): (1971): 1506. doi:10.1001/jama.1971.03180220086028.

Hemilä, Harri, and Elizabeth Chalker. "Vitamin C for preventing and treating the common cold." *Cochrane Database of Systematic Reviews* 1: CD000098 (January 31, 2013). doi:10.1002/14651858.CD000980.pub4.

Jenkins, David J. A., et al. "Supplemental Vitamins and Minerals for CVD Prevention and Treatment." *Journal of the American College of Cardiology* 71(22) (June 2018). jacc.org/doi/full/10.1016/j.jacc.2018.04.020.

Wheeler-Bennett, John W. "Ludendorff: The Soldier and the Politician." *Virginia Quarterly Review* 14(2) (1938): 187-202. jstor.org/stable/i26445446.

5. Smoke without Fire

Skinner, B. F. "Superstition in the pigeon." *Journal of Experimental Psychology* 38(2) (1948): 168-172. doi:10.1037/h0055873.

Goldacre, Ben. *Bad Science*. London: Fourth Estate (2008).

Deer, Brian. "How the case against the MMR vaccine was fixed." *British Medical Journal* 342 (January 6, 2011). doi:10.1136/bmj.c5347.

Godlee, Fiona, Jane Smith, and Harvey Marcovitch. Editorial: "Wakefield's article linking MMR vaccine and autism was fraudulent." *British Medical Journal* 342 (January 6, 2011). doi:10.1136/bmj.c7452.

Andre, F. E., et al. "Vaccination greatly reduces disease, disability, death and inequity worldwide." *Bulletin of the World Health Organization* 86: 140-46 (2008). who.int/bulletin/volumes/86/2/07-040089/en/.

Tversky, Amos, and Daniel Kahneman. "Judgment under Uncertainty: Heuristics and Biases." *Science*, New Series 185(4157) (September 27, 1974): 1124-31. doi:10.1126/science.185.4157.1124.

Kahneman, Daniel. *Thinking, Fast and Slow*. New York: Farrar, Straus and Giroux (2011).

6. The Nature of the Beast

Reuters/Ipsos/UVA Center for Politics Race Poll (September 11, 2017). centerforpolitics.org/crystalball/wp-content/uploads/2017/09/2017-Reuters-UVA-Ipsos-Race-Poll-9-11-2017.pdf.

Canning, David, Sangeeta Raja, and Abdo S. Yazbeck, eds. *Africa's Demographic Transition: Dividend or Disaster?* A copublication of Agence Française de Développement and the World Bank (2015). gbv.de/dms/zbw/81891369X.pdf.

Rutherford, Adam. *A Brief History of EveryoneWho Ever Lived:The Stories in Our Genes*. New York:
The Experiment (2017).

Daley, Tamara C., et al. "IQ on the Rise: The Flynn Effect in Rural Kenyan Children." *Psychological
Science* 14(3) (May 2003), 215–19. jstor.org/stable/40063891.

Ritchie, Stuart. *Intelligence:All that Matters*. London: Hodder & Stoughton (2015).

Bailey, NathanW., and Marlene Zuk. "Same-sex sexual behavior and evolution." *Trends in Ecology
& Evolution* 24(8) (2009): 439–46. doi:10.1016/j.tree.2009.03.014.

Galilei, Galileo. *Dialogue Concerning the Two ChiefWorld Systems: Ptolemaic and Copernican*.
Originally published 1632 as *Dialogo dei massimi sistemi*. Translated and with revised notes
by Stillman Drake. New York: Modern Library Paperback Edition (2001). Published by
arrangement with the University of California Press.

7. Bait and Switch

Darwin, Charles. *On the Origin of Species*. Facsimile of the First Edition: Cambridge,
Massachusetts, and London, England: Harvard University Press (1859). 18th Printing, 2003.

National Academies of Sciences, Engineering, and Medicine. *The Health Effects of Cannabis and
Cannabinoids:The Current State of Evidence and Recommendations for Research*. Washington,
DC: National Academies Press (2017).

Joint Committee on Health. *Report on Scrutiny of the Cannabis for Medicinal Use Regulation Bill
2016*. Dublin: Houses of the Oireachtas (2017). drugsandalcohol.ie/27584/.

8. Schrödinger's Bin Laden

Beale, Geoffrey Herbert. "The cult of T. D. Lysenko: thirty appalling years." *Science Journal*
(October 1969). Quoted in *Toward Dolly: Edinburgh, Roslin and the Birth of Modern
Genetics*. Edinburgh University Centre for Library Collections. Posted January 13,
2014. libraryblogs.is.ed.ac.uk/towarddolly/2014/01/13/
the-lysenko-controversy-soviet-genetics-and-edinburgh/.

Festinger, Leon, et al. *When Prophecy Fails:A Social and Psychological Study of a Modern GroupThat
Predicted the Destruction of theWorld*, Harper-Torchbooks (1956). Reprinted 2009 by Martino
Fine Books, Eastford, CT.

Allen, M. R., et al. Intergovernmental Panel on Climate Change (IPCC). Climate Change 2014
Synthesis Report Fifth Assessment Report). ar5-syr.ipcc.ch.

Diethelm, Pascal, and Martin McKee. "Denialism:What is it and how should scientists respond?"
European Journal of Public Health 19(1) (January 2009): 2-4. doi:10.1093/eurpub/ckn139.

Weart, Spencer. "Global warming: How skepticism became denial. *Bulletin of the Atomic Scientists*
67(1) (January 1, 2011): 41-50. thebulletin.org/2011/01/global-warming-how-skepticis
m-became denial/.

Lewandowsky, Stephan, et al. "NASA Faked the Moon Landing—Therefore, (Climate) Science
Is a Hoax:An Anatomy of the Motivated Rejection of Science." *Psychological Science* 24(5)
(2013): 622-33. doi:10.1177/0956797612457686.

Grimes, David Robert. "Denying climate change isn't scepticism—it's 'motivated reasoning'"
The Guardian (February 5, 2014). theguardian.com/science/2014/feb/05/
denying-climate-change-scepticism-motivated-reasoning.

Kahan, Dan M., et al. "Motivated numeracy and enlightened self-government." *Behavioral Public
Policy* 1(1) (2017): 54-86. doi:10.1017/bpp.2016.2.

9. The Memory Remains

National Research Council. Identifying the Culprit:Assessing Eyewitness Identification.
Washington, DC: National Academies Press (2015).

Sacks, Oliver. "Speak, Memory." *New York Review of Books* (February 21, 2013).
Loftus, E. F., and J. E. Pickrell. "The Formation of False Memories." *Psychiatric Annals* 25(12) (1995): 720–25. doi:10.3928/0048-5713-19951201-07.
Loftus, Elizabeth F. "Planting misinformation in the human mind: a 30-year investigation of the malleability of memory." *Learning & Memory* 12(4) (July–August 2005): 361–66. doi:10.1101/lm.94705.
Schreiber, Nadja, et al. "Suggestive interviewing in the McMartin Preschool and Kelly Michaels daycare abuse cases: A case study." *Social Influence* 1(1) (2006): 16–47. doi:10.1080/15534510500361739.

10. Daggers of the Mind

Moore, Timothy E. "Scientific Consensus and Expert Testimony: Lessons from the Judas Priest Trial." *Skeptical Inquirer* 20(6) (November/December 1996). skepticalinquirer.org/1996/11/scientific-consensus-and-expert-testimony-lessons-from-the-judas-priest-trial/.
Blanke, Olaf, et al. "Neurological and robot-controlled induction of an apparition." *Current Biology* 24(22) (November 17, 2014): 2681–86. doi:10.1016/j.cub.2014.09.049.
Cheyne, J. A., et al. "Hypnagogic and hypnopompic hallucinations during sleep paralysis: neurological and cultural construction of the night-mare." *Consciousness and Cognition* 8(3) (September 1999): 319–37. doi:10.1006/ccog.1999.0404.
Chevreul, Michel Eugène. "De la baguette divinatoire: du pendule dit explorateur et des tables tournantes, au point de vue de l'histoire de la critique et de la méthode expérimentale." Mallet-Bachelier (1854). books.google.com/books/about/De_la_baguette_divinatoire.html?id=X5bF8qCWJnQC.
Mercier, C. A. "Automatic Writing." *British Medical Journal* 1(1726) (January 27, 1894): 198–99. www.ncbi.nim.nih.gov/pmc/articles/PMC2403845/.
Mostert, Mark P. "An activist approach to debunking FC." *Research and Practice for Persons with Severe Disabilities* (2014): 203–10. doi:10.1177/1540796914556779.
Wheeler, D. L, et al. "An experimental assessment of facilitated communication." *Mental Retardation* 31(1) (1993): 49. pubmed.ncbi.nim.nih.gov/8441353/.
Mostert, Mark P. "Facilitated Communication and Its Legitimacy—Twenty-First Century Developments." *Exceptionality* 18(1) (2010): 31–41. doi:10.1080/09362830903462524.

11. Great Expectations

Forer, B. R. "The fallacy of personal validation: a classroom demonstration of gullibility." *Journal of Abnormal Psychology* 44(1) (1949): 118. doi:10.1037/h0059240.
Carlson, Shawn. "A double blind test of astrology." *Nature* 318 (1985): 419–25. doi:10.1038/318419a0.
Pittenger D. J., "Measuring the MBTI . . . and coming up short." *Journal of Career Planning and Employment* 54(1) (January 1993): 48–52. researchgate.net/publication/237675975_Measuring_the_MBTI_and_coming_up_short.
Montgomery, Guy, and Irving Kirsch. "Mechanisms of Placebo Pain Reduction: An Empirical Investigation." *Psychological Science* 7(3) (1996): 174–76. doi:10.1111/j.1467-9280.1996.tb00352.x.
Ernst, E. "The attitude against immunisation within some branches of complementary medicine." *European Journal of Pediatrics* 156(7) (1997): 513–15. doi:10.1007/s004310050650.
Rubin, G. James, R. Rosa Nieto-Hernandez, and Simon Wessely. "Idiopathic environmental intolerance attributed to electromagnetic fields (formerly 'electromagnetic hypersensitivity'): An updated systematic review of provocation studies." *Bioelectromagnetics* 31(1) (2010): 1–11. doi:10.1002/bem.20536.
World Health Organization. "Electromagnetic fields and public health: Electromagnetic hypersensitivity." (December 2005). who.int/teams/environment-climate-change-and-health/radiation-and-health/electromagnetic-fields-and-public-health-electromagnetic-hypersensitivity.

Lamberg, M., H. Hausen, and T. Vartiainen. "Symptoms experienced during periods of actual and
   supposed water fluoridation." *Community Dentistry and Oral Epidemiology* 25(4) (August 1997):
   291-95. doi:10.1111/j.1600-0528.1997.tb00942.x.
Kruger, J. and D. Dunning. "Unskilled and unaware of it: how difficulties in recognizing one's own
   incompetence lead to inflated self-assessments." *Journal of Personality and Social Psychology*,
   77(6) (December 1999): 1121-34. doi:10.1037/0022-3514.77.6.1121.

12. Chance Encounters

Selvin, Steve, et al. "A Problem in Probability (Letter to the Editor)." *American Statistician* 29(1)
   (1975): 67. jstor.org/stable/2683689.
Herbranson, Walter T., and Julia Schroeder. "Are Birds Smarter Than Mathematicians? Pigeons
   (*Columba livia*) Perform Optimally on a Version of the Monty Hall Dilemma." *Journal of
   Comparative Psychology* 124(1) (2010): 1-13. doi:10.1037/a0017703.
Gigerenzer, Gerd. *Reckoning with Risk: Learning to Live with Uncertainty.* London: Penguin
   UK (2003).
Royal Statistical Society. "Royal Statistical Society concerned by issues raised in Sally Clark case."
   News Release, October 23, 2001.inference.org.uk/sallyclark/RSS.html.
Royal Statistical Society. "Letter from the President to the Lord Chancellor regarding the
   use of statistical evidence in court cases." (2002). Cited in Richard Nobles and David
   Schiff, "Misleading statistics within criminal trials." *Significance*, 2(1) February 24, 2005.
   doi:10.1111/j.1740-9713.2005.00078.x.
Watkins, S. J. "Conviction by mathematical error? Doctors and lawyers should get probability
   theory right." *British Medical Journal* (2000): 2-3. doi:10.1136/bmj.320.7226.2.

13. Sifting the Signal

Bickel, P. J., E. A. Hammel, and J. W. O'Connell. "Sex bias in graduate admissions: Data from
   Berkeley." *Science* 187(4175) (1975): 398-404. doi:10.1126/science.187.4175.398.
Appleton, David R., Joyce M. French, and Mark P. J. Vanderpump, "Ignoring a covariate: An
   example of Simpson's paradox." *American Statistician* 50(4) (1996): 340-1. doi:10.1080/
   00031305.1996.10473563.
Vigen, Tyler. *Spurious Correlations.* New York: Hachette Books (2015).
Ioannidis, John P. A. "Stealth Research: Is Biomedical Innovation Happening Outside the
   Peer-Reviewed Literature?" *Journal of the American Medical Association* 313(7) (2015): 663-64.
   doi:10.1001/jama.2014.17662.
Diamandis, Eleftherios P. "Theranos phenomenon: promises and fallacies." *Clinical Chemistry and
   Laboratory Medicine* 53(7) (2015): 989-93. doi:10.1515/cclm-2015-0356.

14. Size Matters

Yong, Ed. "Beefing with the World Health Organization's Cancer Warnings." *The Atlantic* (October
   26, 2015). theatlantic.com/health/archive/2015/10/
   why-is-the-world-health-organization-so-bad-at-communicating-cancer-risk/412468/.
Ioannidis, John P. A. "Why Most Published Research Findings Are False." *PLoS Medicine* 2(8) (2005):
   e124. doi:10.1371/journal.pmed.0020124
Colquhoun, David. "An investigation of the false discovery rate and the misinterpretation of
   p-values." *Royal Society Open Science* 1(3) (2014): 140216. doi:10.1098/rsos.140216.
Grimes, David Robert, Chris T. Bauch, and J. P. A. Ioannidis. "Modelling science trustworthiness
   under publish or perish pressure." *Royal Society Open Science* 5(1) (2018): 171511. doi:10.1098/
   rsos171511.

15. Skewing the Balance

Krugman, Paul. "The Falsity of False Equivalence." *The New York Times* (September 26, 2016). krugman.blogs.nytimes.com/2016/09/26/the-falsity-of-false-equivalence/.

Grimes, David Robert. "Impartial journalism is laudable. But false balance is dangerous." *The Guardian* (November 8, 2016). theguardian.com/science/blog/2016/nov/08/impartial-journalism-is-laudable-but-false-balance-is-dangerous.

Grimes, David Robert. "A dangerous balancing act: On matters of science, a well-meaning desire to present all views equally can be a Trojan horse for damaging falsehoods." *EMBO Reports* 20(8) (July 8, 2019): e48706. doi:10.15252/embr.201948706.

Michaels, David. "Doubt Is Their Product." *Scientific American*, 292(6) (2005): 96–101. scientificamerican.com/article/doubt-is-their-product/.

Boykoff, Maxwell T., and Jules. M. Boykoff. "Balance as bias: global warming and the US prestige press." *Global Environmental Change* 14(2) (2004): 125–36. eci.ox.ac.uk/publications/downloads/boykoff04-gec.pdf.

BBC Trust. *Trust Conclusions on the Executive Report on Science Impartiality Review Actions.* (July 2014). downloads.bbc.co.uk/bbctrust/assets/files/pdf/our-work/science_impartiality/trust_conclusions.pdf.

Brüggemann, Michael, and Sven Engesser. "Beyond false balance: How interpretive journalism shapes media coverage of climate change." *Global Environmental Change* 42 (January 2017): 58–67. doi:10.1016/j.gloenvcha.2016.11.004.

16. Tales from the Echo Chamber

Bakshy, Eytan, Solomon Messing, and Lada A. Adamic. "Exposure to ideologically diverse news and opinion on Facebook." *Science* 348(6239) (June 5, 2015): 1130–32. doi:10.1126/science.aaa1160.

Del Vicario, Michela, et al. "The spreading of misinformation online." *PNAS* 113(3) (2016): 554–59. doi:10.1073/pnas.1517444113.

van Alstyne, Marshall, and Erik Brynjolfsson. "Electronic Communities: Global Villages or Cyberbalkanization? (Best Theme Paper)." *ICIS 1996 Proceedings* (1996): 5. aisel.aisnet.org/icis1996/5.

Gandour, Ricardo. "Study: Decline of traditional media feeds polarization." *Columbia Journalism Review* (September 19, 2016). cjr.org/analysis/media_polarization_journalism.php.

Maddox, John. "Has Duesberg a right of reply?" *Nature* 363 (1993): 109. doi:10.1038/363109a0.

17. The Outrage Machine

Williamson, Elizabeth. "Truth in a Post-Truth Era: Sandy Hook Families Sue Alex Jones, Conspiracy Theorist." *The New York Times* (May 23, 2018). nytimes.com/2018/05/23/us/politics/alex-jones-trump-sandy-hook.html

Silverman, Craig. "This Is How Your Hyperpartisan Political News Gets Made." BuzzFeed News (February 27, 2017). buzzfeednews.com/article/craigsilverman/how-the-hyperpartisan-sausage-is-made.

Grimes, David Robert. "Russian fake news is not new: Soviet Aids propaganda cost countless lives." *The Guardian* (2017). theguardian.com/science/blog/2017/jun/14/russian-fake-news-is-not-new-soviet-aids-propaganda-cost-countless-lives.

Andrew, Christopher, and Vasili Mitrokhin. *The Sword and the Shield: The Mitrokhin Archive and the Secret History of the KGB*, Hachette UK (2000).

United States Department of State. "Soviet Influence Activities: A Report on Active Measures and Propaganda, 1986-87." (August 1987). globalsecurity.org/intell/library/reports/1987/soviet influence-activities-1987.pdf.

Barnes, Julian E., Matthew Rosenberg, and Edward Wong. "As Virus Spreads, China and Russia
See Openings for Disinformation." *The New York Times* (March 28, 2020; updated October 6,
2020). nytimes.com/2020/03/28/us/politics/china-russia-coronavirus-disinformation.html.

18. Bad Influencers

Donelly, Beau, and Nick Toscano. *The Woman Who Fooled the World: Belle Gibson's Cancer Con, and the
Darkness at the Heart of the Wellness Industry.* Scribe US (2018).
Grimes, David Robert. "Beware the snake-oil merchants of alternative medicine–your life could
depend on it." *The Irish Times* (March 6, 2018). irishtimes.com/life-and-style/health-family/
beware-the-snake-oil-merchants-of-alternative-medicine-your-life-could-depend-on
-it-1.3405985.
Lancucki, L. et al. "The impact of Jade Goody's diagnosis and death on the NHS Cervical Screening
Programme." *Journal of Medical Screening* 19(2) (2012): 89–93. doi:10.1258/jms.2012.012028.
Cocozza, Paula. "Whatever happened to the Jade Goody effect?" *The Guardian* (January 21,
2018). theguardian.com/society/shortcuts/2018/jan/21/whatever-happened-to-the-jad
e-goody-effect.
Chapman, Simon, et al. "Impact of news of celebrity illness on breast cancer screening: Kylie
Minogue's breast cancer diagnosis." *Medical Journal of Australia* 183(5) (2005): 247–50.
doi:10.5694/j.1326-5377.2005.tb07029.x.
Gorski, David. "The Oprah-fication of medicine." *Science-Based Medicine* (June 1, 2009).
sciencebasedmedicine.org/the-oprah-fication-of-medicine/.
Korownyk, Christina, et al. "Televised medical talk shows—what they recommend and the
evidence to support their recommendations: a prospective observational study." *British
Medical Journal* 349 (2014): g7346. doi:10.1136/bmj.g7346.
Gunter, Jennifer. *The Vagina Bible: The Vulva and the Vagina: Separating the Myth from the Medicine.*
Citadel (2019).
Pennycook, Gordon, et al. "On the reception and detection of pseudo-profound bullshit." *Judgment
and Decision Making* 10(6) (November 2015): 549–63. journal.sjdm.org/15/15923a/jdm15923a.pdf.

19. The Edge of Science

Grimes, David Robert. "Proposed mechanisms for homeopathy are physically impossible," *Focus
on Alternative and Complementary Therapies* 17(3) (2012): 149–55. doi:10.1111/j.2042-7166-2012.01162.x.
Maddox, John, James Randi, and Walter W. Stewart. "High-dilution' experiments a delusion."
Nature 334(6180) (1988): 287–90. doi:10.1038/334287a0.
Sagan, Carl. *The Demon-Haunted World: Science as a Candle in the Dark.* New York: Random
House (1995).
England, Philip C., Peter Molnar, and Frank M. Richter. "Kelvin, Perry and the Age of the Earth."
American Scientist 95(4) (2007): 342–49. americanscientist.org/article/kelvin-perry-and-th
e-age-of-the-earth.
Popper, Karl. "The Logic of Scientific Discovery" (1959). Reprinted 2014 by Martino Fine Books,
Eastford, CT.

20. Rise of the Cargo Cult

Feynman, Richard P. "Cargo Cult Science." California Institute of Technology commencement
address (1974).
The Irish Expert Body on Fluorides and Health, Appraisal of "Human toxicity, environmental
impact and legal implications of water fluoridation." (2012). fluoridesandhealth.ie/assets/files/
documents/Appraisal_of_Waugh_report_May_2012.pdf.
Committee on Identifying the Needs of the Forensic Sciences Community, National Research
Council. "Strengthening Forensic Science in the United States: A Path Forward." National
Academies Press (August 2009). ncjrs.gov/pdffiles1/nij/grants/228091.pdf.

Federal Bureau of Investigation. "FBI Testimony on Microscopic Hair Analysis Contained Errors in at Least 90 Percent of Cases in Ongoing Review." (2015). fbi.gov/news/pressrel/ press-releases/fbi-testimony-on-microscopic-hair-analysis-contained-errors-in-at-least-90-percent-of-cases-in-ongoing-review.

21. A Healthy Skepticism

Oliver, J. Eric, and Thomas Wood. "Medical conspiracy theories and health behaviors in the United States." *JAMA Internal Medicine* 174(5) (2014): 817-18. doi:10.1001/ jamainternmed.2014.190.

Grimes, David Robert. "Six stubborn myths about cancer." *The Guardian* (August 30, 2013). theguardian.com/science/2013/aug/30/six-stubborn-myths-cancer.

Johnson, Skyler B., et al. "Complementary Medicine, Refusal of Conventional Cancer Therapy, and Survival among Patients with Curable Cancers." *JAMA Oncology* 4(10) (2018): 1375-81. doi:10.1001/jamaoncol.2018.2487.

United Nations Scientific Committee on the Effects of Atomic Radiation. UNSCEAR 2008 Report Vol. II. Sources and Effects of Ionizing Radiation. Annex D: Health effects due to radiation from the Chernobyl accident. (2011). unscear.org/unscear/en/publications/2008_2.html.

World Health Organization. "Health effects of the Chernobyl accident and special health care programmes: an overview." (2006). who.int/publications/i/item/9241594179.

Grimes, David Robert. "Why it's time to dispel the myths about nuclear power." *The Guardian* (April 11, 2016). theguardian.com/science/blog/2016/apr/11/ time-dispel-myths-about-nuclear-power-chernobyl-fukushima.

Swami, Viren, et al. "Analytic thinking reduces belief in conspiracy theories." *Cognition* 133(3) (2014): 572-85. doi:10.1016/j.cognition.2014.08.006.

Epilogue

Nyhan, Brendan, et al. "Effective messages in vaccine promotion: a randomized trial." *Pediatrics* 133(4) (2014): e835-42. doi:10.1542/peds.2013-2365.

Jolley, Daniel, and Karen M. Douglas. "The Effects of Anti-Vaccine Conspiracy Theories on Vaccination Intentions." *PloS One* 9(2) (February 2014). doi:10.1371/journal.pone. 0089177.

Prue, Gillian, et al. "Access to HPV vaccination for boys in the United Kingdom." *Medicine Access @ Point of Care.* (September 2018). doi:10.1177/2399202618799691.

Corcoran, Brenda, Anna Clarke, and Tom Barrett. "Rapid response to HPV vaccination crisis in Ireland." *The Lancet* 391(10135) (May 28, 2018): 2103. doi:10.1016/S0140-6736(18)30854-7.

Grimes, David Robert. "Anti-HPV vaccine myths have fatal consequences." *The Irish Times* (September 8, 2017). irishtimes.com/opinion/antyi-hpv-vaccine-myths-hav e-fatal-consequences-1.3213118.

Mitchell, Susan. "Regret's regrettable behaviour." *Business Post* (July 23, 2017). businesspost.ie/ insight/susan-mitchell-regrets-regrettable-behaviour-3164a1eo.

Grimes, David Robert. "Terminally Ill at 25 and Fighting Fake News on Vaccines." *The New York Times* (December 11, 2019). nytimes.com/2019/12/11/opinion/anti-vaccine-HPV.html.

Imhoff, Roland, and Pia Karoline Lamberty. "Too special to be duped: Need for uniqueness motivates conspiracy beliefs." *European Journal of Social Psychology*, 47(6) (2017): 724-34. doi:10.1002/ejsp.2265.

Crotty, David. "The Guardian Reveals an Important Truth About Article Comments." Society for Scholarly Publishing: *Scholarly Kitchen* (January 4, 2013). scholarlykitchen.sspnet.org/ 2013/01/04/the-guardian-reveals-an-important-truth-about-article-comments/.

Popper, Karl R. *The Open Society and Its Enemies*, Vol. 1, The Spell of Plato. London: Routledge. 1st ed. 1945; First single-volume Princeton University Press printing, 2013; Princeton Classics paperback edition, first printing, 2020.

ACKNOWLEDGMENTS

Very little worthwhile ever happens in isolation. The best science is collaborative, and in many respects so is a book like this—it is only possible due to the expertise and support of a diverse array of people.

The stories herein are underpinned by a litany of scientific, medical, and psychological research, stemming from the efforts of multitudinous researchers. I am immensely grateful for these insights and hope I've done justice in communicating these findings. Where appropriate, I've cited the original literature and related material for the interested reader.

I owe debts of gratitude to my agent, Patrick Walsh, John Ash, and Brian Langan for their invaluable guidance, and to my UK editor, Ian Marshall. I especially want to thank my editorial collaborators on this edition, Nicholas Cizek and Liana Willis, both of whom were instrumental in the evolution of this updated and expanded text. I thank Richard Dawkins for his very helpful insights and support, and Robin Ince for his kindness throughout the process. I especially wish to thank Simon Singh, for without his support it's highly unlikely I would have ever had the impetus to start.

I thank also my various editors at outlets who've given me the opportunity to write and broadcast, particularly *The Irish Times*, *The Guardian*, the *Sunday Business Post,* and the BBC. I am grateful too to Sense About Science, who recognized my propensity for getting myself in trouble early on, and whose recognition convinced me to continue advocating for science even in adverse conditions. Because often, adverse conditions are when such advocacy is most required.

Over the years I've been blessed with phenomenal academic support, superb colleagues, and fantastic mentors. I'd like to extend my deep appreciation to Enda McGlynn, David Basanta, and Mike Partridge in particular—all of whom took chances on me as a young scientist, shaping both my career and my outlook on what scientists can offer the world. I thank, too, the universities with which I've been affiliated—the University of Oxford, Queen's University Belfast, and Dublin City University—both for the wonderful research opportunities and the support on occasions when my work made me a target.

I've been lucky to have a fantastically supportive network of science-minded friends and colleagues whose support has been invaluable: Robert O'Connor, Anthony Warner, Susan Mitchell, Dorothy Bishop, Ciara Kelly, Daniela Robles, Eileen O'Sullivan, Donal Brennan, David Colquhoun, David Gorski, Christopher French, the Brennan family, and so many others I have neglected to mention due to brevity or thoughtlessness.

It is those closest to me, though, who have always supported me, sculpting my ideas and forgiving my errors and excesses. While there are far too many to mention specifically, I need to thank Mathilde Hernu especially, as well as Danny Murray, Laura Brennan, and Graham Keatley for more than I can ever express. Last but most important, I owe everything to my family: my fantastic brother Stephen and my incredible parents, Patricia and Brendan. They have shaped the man I am today—without them, this book wouldn't exist.

INDEX

Page numbers in *italic* refer to photographs, illustrations, and tables; n indicates footnotes.

ABOUT THE AUTHOR

DR. DAVID ROBERT GRIMES is a physicist, cancer researcher and science journalist. Born in Dublin in 1985, he is affiliated with Dublin City University and University of Oxford. He contributes to both the BBC and RTE discussing science, politics and media and has contributed to *The Guardian*, *The Irish Times*, the BBC, PBS, and *The New York Times*, among others. He also advises on science policy, and was joint recipient of the 2014 Nature/Sense about Science Maddox Prize for Standing Up for Science.

davidrobertgrimes.com   🐦drg1985   @david_robert_grimes